PHYSICS 1

LABORATORY MANUAL

With PASCO Capstone Supporting Software

John Winfrey

Upon adoption, a copy of the Capstone Software *specifically designed for these labs* is available from:

physicsprof6022@yahoo.com

Site License $125

You also need a Capstone License from PASCO.

** Professional photos by permission of 123RF.com; Shutterstock.com; and The American Physical Society

Contents

PREFACE

For Students & Instructors

In general, Introductory Physics Laboratories provide the contents and procedures expected in most science laboratory courses.

However, a ***robust*** Physics Laboratory Course contains, in addition, several subliminal elements.

Laboratory Experiments Should:

1. Provide experience with making measurements and following scientific procedures. When a fork in the road arises, the lab manual should promote exercising the scientific method to make decisions.
2. Provide tactile support for Lecture course materials, especially with abstract or microscopic and macroscopic concepts beyond the given human sense. Kinesthetic learning (Muscle memory).
3. Provide concrete, real-world experiences manipulating the Mathematics and Physics employed in the Lecture course.
4. Provide guided explorations as well as by-inquiry explorations.
5. Promote equal time for making measurements and drawing conclusions. Students should obtain coherent understanding of core concepts presented.
6. Be flexible enough for either 2-hr. or 3-hr. classes.
7. Contain a sufficiently stand-alone introduction for students to cope with the frequent mismatch between the Laboratory sequence and the Lecture sequence.
8. Seriously address common misconceptions.

This manual will be based upon Pasco's ***Capstone*** data acquisition software. The student "Workbooks" for this software are contained in a digital file, with all data measurements seen on the computer screen embedded in the file for printing or for saving an eVersion for themselves for future reference and review.

This Lab Manual will be basically a "wrapper" for a ***Capstone*** suite of experiments.

Experiment Elements

- A stand-alone lesson introducing the relevant Physics topic to be explored
- Electronic Student Workbooks
- Pre-laboratory thought questions
- Post-lab re-evaluation of pre-laboratory questions
- Sometimes, an interactive group discussion concentrating on the core principles that were explored.

The Issue of Misconceptions

I tell my lecture students the first meeting time that about 50% of what they *think* they know about how nature works is *wrong*! Note: when I have truly international students, these misconceptions are mostly absent. So the problem is mostly Americana.

These misconceptions arrive in students minds through a variety of known and unknown means. A few of the culprits that are known are:

a. Early explanations to *supposedly* immature children that are watered down. Like the Stork story of child birth.
b. Early explanations to children by adults who are not humble enough to say, "I don't know, let's go look this up", instead of a face-saving, plausible untruth.
c. Television. Blurring between reality and fantasy and special effects.
d. Errors that have crept into our text books, and teacher preparation materials. Many public school teachers convey wrong information, in all good faith.

Furthermore, these misconceptions, by the time students arrive in Postsecondary Education, are deeply buried and defended by the subconscious.

Merely informing a student that they have a misconception causes ego-defense. And, unfortunately, adding correct information on top of misconception results in a reversion to the misconception again a few months down the road.

There have been great efforts by Physics Education Research (PER) groups to solve this problem. The results:

I.   We have extensive lists of those misconceptions in Physics[1].

"Misconceptions (a.k.a. alternative conceptions, alternative frameworks, etc.) are a key issue from Constructivism in science education, a major theoretical perspective informing science teaching. In general, **scientific misconceptions** have their foundations in a few intuitive knowledge domains, including folk mechanics (object boundaries and movements), folk biology (biological species configurations and relationships), and folk psychology (interactive agents and goal-directed behavior). This enables humans to interact effectively with the world in which they evolved. That these folk sciences do not map accurately onto modern scientific theory is not unexpected. A second major source of scientific misconceptions are instruction-induced or didaskalogenic misconceptions."[2]

---

[1] http://amasci.com/miscon/opphys.html
[2] Wikipedia.

II.   We have an assessment instrument[3] normed on 12,000 students (the Force Concepts Inventory, FCI).

III.   Research has provided a methodology that roots out these incorrect concepts, and replaces them with correct concepts:

For each misconception, we will employ the CCM model[4]

   a. Ask students questions about what they feel in their guts is true about a relevant concept in the pre-lab, and force them to write down their misconceptions. Without this commitment, students will misremember what they used to think; the mind doesn't like to acknowledge errors.

   b. Confront students directly and brutally with contrary evidence through measurement. Pasco **Capstone** experiments.

   c. Ask students to revisit their prior conceptions and their measurements, and create a new reality.

***

Student's scores on Lecture class examinations have been shown to directly correlate with their success in this reprogramming. Students' FCI post-test score predicted about 45% of the variance in course grades.

Example: Suppose a student has a misconception about what the path of a package dropped from an aircraft is. Straight down or parabolic (having the same initial horizontal velocity as the aicraft at separation. Then the student will draw an incorrect master diagram, do (probably) the correct mathematics for the incorrect path, and still miss the answer.

It is especially crucial for these misconceptions to be eradicated from student's subconscious minds before taking the GRE or any of the xxxCAT's.

---

[3] Hake (1998). *Interactive-Engagement Versus Traditional Methods: A Six-Thousand-Student Survey Of Mechanics Test Data For Introductory Physics Courses*. American Journal of Physics, 66, 64-74.

[4] *Evaluation Novelty in Modeling-Based and Interactive Engagement Instruction, Eurasia Journal of Mathematics, Science & Technology Education*, 2007, **3**(3), 231-237.

Lab 1: Kinematics: One Dimensional Motion

Name _____

Read the Preface before attempting this the first time!

Please respond to the following questions without reading ahead in the Lab book or your Text book. Like free-association, just put down on paper what your gut-level response is:

Psychologists define "learning" as a relatively permanent change in behavior based upon experiences.

1. According to that definition, I can learn only by observing my environment.

2. Graphs and Words and Mathematics are for describing different kinds of ideas.

Lab 1

Kinematics: One Dimensional Motion

Equipment:

Motion Detector
Computer Interface

We are ahead of lecture, trying to integrate three separate ways of analyzing and understanding and categorize motion. Soon, you will need to be able to integrate the three following methods of description into one complete, integrated thought:

> **English Description** (some of which may not match Physics usage)
> **Graphs**
> **Equations** that describe this motion[5]

Today: We will work on "At rest" and "Uniform Motion" (speedometer stays constant, like cruise control). Next week: "Accelerated" motion: the speedometer is increasing or deceasing

A. Graphs

This week we will be associating the graph of x vs. t (called a World Line) with English terminology. The figures below are position vs. time physically and graphically in 1 D, .

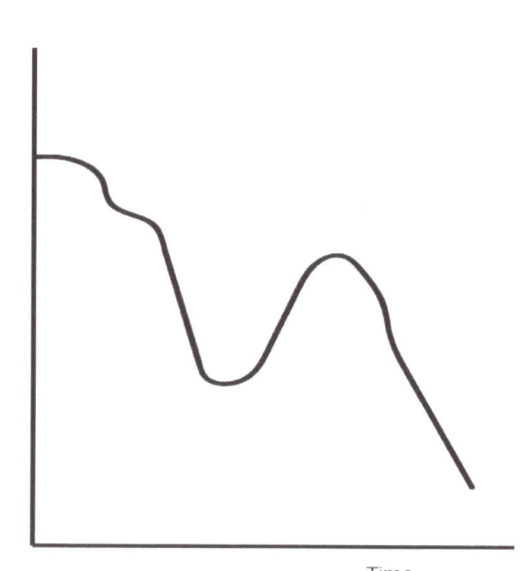

---

[5] This will be covered extensively in lecture, and in later Labs.

"In Physics, the **world line** of an object is the unique path of that object as it travels through 4-dimensional space-time. The idea of world lines originates in physics and was pioneered by Hermann Minkowski ….

"… world lines are a general way of representing the course of events. The use of it is not bound to any specific theory. Thus in general usage, a world line is the sequential path of personal human events (with *time* and *place* as dimensions) that marks the history of a person — perhaps starting at the time and place of one's birth until one's death."[6]

A graph is a visual display format that shows (or doesn't show: see chaos theory) the connection between two variables. Mathematicians, *by convention*, plot the first two (dummy) dimensions as y vs. x.

Physicists are often forgetful and easily confused, so we try to give the dummy variables names (often a single letter) that give a *hint* as to what the variable means in the laboratory, i.e., x, y, z, t.

✓ **Position**

Thus, one of todays' graphs will be of x (in particular, "your" position relative to a spatial origin: the motion detector) vs. t (real time since you started the measurement). You will be moving so as to match a given graph.

---

[6] Wikipedia.

Motion Detector on Left Margin

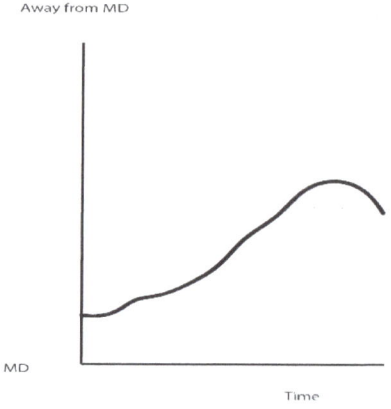

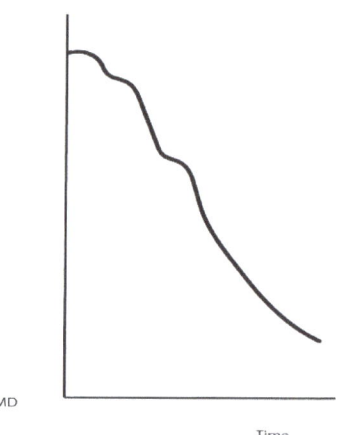

Figure 1-3: The left figures (a and b) and the right figures (c and d) are different descriptions of the <u>same</u> set of events through time. Away from detector is positive; toward detector negative.

In this experiment, you will be learning both the connection between the graph and your corresponding movement, and also how "you" can "control" motion to form a specific graph (word description).

For example, what do you **do** to get to your new person-of-interest's house using a map?

    ✓ **Kinesthetic Learning**

Kinesthetic learning (also known as **tactile learning**) is a learning style in which learning takes place by the student carrying out a physical activity, rather than listening to a lecture or watching a demonstration. People with a preference for kinesthetic learning are also commonly known as "do-ers".[7]

---

[7] Wikipedia

Skilled surgeons practice so many times, that their "cuts" are an automatic muscle control sequences: they acquire kinesthetic knowledge.

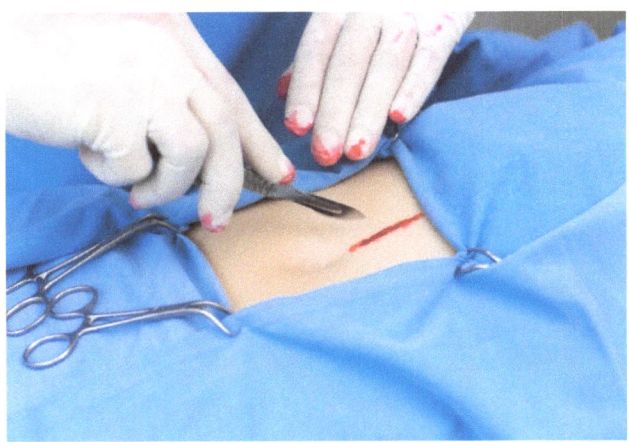

Fig 1-4 Kinesthetic learning.

"**Constructivism**, a perspective in education, which is based upon experiments on learning. In the process of learning, real life experience is used to construct and conditionalize current new experiences (developing a schema to explain that experience)."[8]

It is problem based, adaptive learning that challenges <u>faulty schema</u>[9], integrates new knowledge with existing knowledge, and allows for creation of original work or innovative procedures. These types of learners are self-directed, creative, innovative, drawing upon visual/spatial, musical/rhythmic, bodily kinesthetic, verbal/linguistic, logical/mathematical, interpersonal, intrapersonal, and naturalistic intelligences. The purpose in education is to become creative and innovative through analysis, conceptualizations, and synthesis of prior experience to create *new* knowledge." [10]

✓ **Velocity or Speed**

The velocity and the speed of a moving object is the rate with which the object moves per unit time.

Suppose you are on a trip from Huntington to Chicago. There are several paths you could take to accomplish the same change in position, but which would take different amounts of time to complete. Is your issue "time on trip" or "diversity on trip"?

---

[8] Wikipedia.

[9] A mental codification of experience that includes a particular organized way of perceiving cognitively and responding to a complex situation or set of stimuli

[10] Wikipedia.

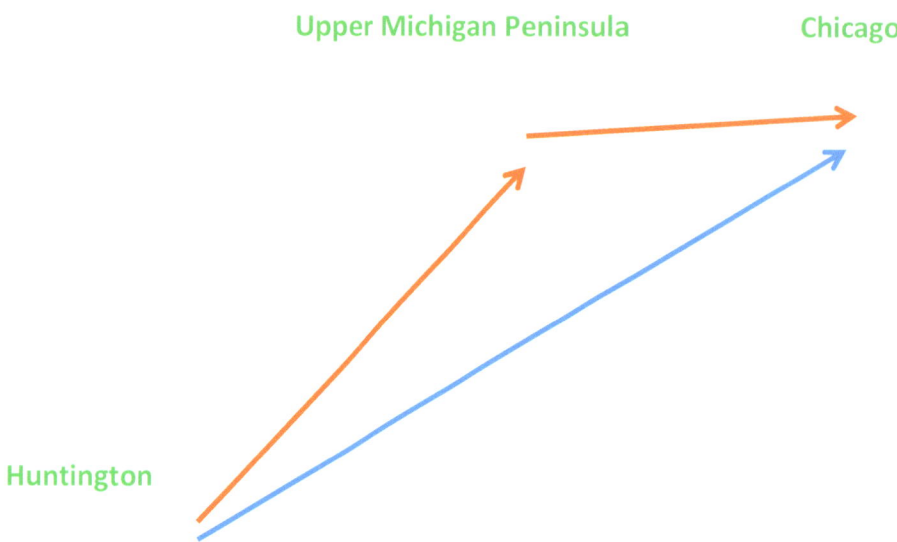

Fig 1-5 Alternate Ways to Travel to Chicago.

A way to distinguish the identical, total-outcome trips (changes in position as the crow flies) is to define your **average speed**:

$$\bar{v} = \text{change in odometer/total-time}$$

The orange and blue lines have differences here.

Or your **average velocity** which is

$$v_{avg} = \text{total displacement[11]/total time}$$

The orange and blue lines have the same value here.

Or your **instantaneous speed** in various short term time intervals (what your speedometer does):

$$\text{instantaneous speed} = \text{speed within a short time period[12]}$$

The latter distinguishes the exact details of a particular total trip.

Which one you use depends upon if you are interested in the net outcome, or the detailed description.

In the second part of this Lab, you will be moving so as to match position and velocity graphs.

Notice the difference in graphics between a slow velocity and a fast velocity (speed) as you experiment with this parameter.

---

[11] Straight line distance between departure and arrival points.
[12] What your seedometer does.

Fig 1-6 Cartoons of a) Slow and b) Fast

B. Physics Terms (do not always match common English usage): The beginning of a list!

> STOP
> STEADILY (AT CONSTANT SPEED)
> SLOW (RATE)
> FAST (RATE)
> TOWARD (DIRECTION)
> AWAY (DIRECTION)
> SLOWING DOWN[13]

THERE IS A RELATIONSHIP, GRAPHICALLY, BETWEEN (POSITION VS. TIME) AND (VELOCITY VS. TIME) THAT YOU WILL NEED TO DISCOVER.

Please use the "Journal Option" while taking data:

1. So that if you erase or lose a previous data sample or your computer freezes, you will have a record.
2. You can email these electronic Workbooks to yourself for studying, or save them to a flash drive or to your "student" drive. These are useful for both the Lecture and the Lab portions of the course.

> ➤ **Open Capstone Workbook 1 and follow instructions. Journal.**

---

[13] Physics technical definition doesn't match common English Usage; we will tackle this one later

Lab 1

Kinematics: One Dimensional Motion

Post-Lab Concept Issues:

Revisit the Pre-Lab questions.

For questions where your thinking has not change, mark NC.

For questions where the experiment has changed your understanding, re-answer the question as you now understand it.

Psychologists define "learning" as a relatively permanent change in behavior based upon experiences.

1. According to that definition, I can learn only by observing my environment.

2. Graphs and words and mathematics are for describing different kinds of ideas.

Lab 2: Acceleration

Name _____

<u>Pre-Lab Concept Issues</u>:

Please respond to the following questions without reading ahead in the Lab book or your Text book. Like free-association, just put down on paper what your gut-level response is:

1. The terms distance and displacement are synonymous and may be used interchangeably. Thus the distance an object travels and its displacement are always the same.

2. Velocity is another word for speed. An object's speed and velocity are always the same.

3. Acceleration and speed are the same thing.

4. Acceleration always means that an object is speeding up.

5. Acceleration always occurs in the same direction as an object is moving.

6. If an object has a speed of zero (even instantaneously), it has no acceleration.

Lab 2 Acceleration

Equipment:

Motion sensor
Ceiling bracket for Motion sensor
Ball
Fan cart
2.0 m Dynamics track with motion sensor attached

Introduction:

This week we will be building upon the concepts of (position vs. time) and (velocity vs. time) from the previous Lab.

Last week, we restricted ourselves to "uniform" (constant) velocity. That was enough to build a foundation.

So, what if we relax that condition? We allow ***accelerated*** motion.

For your vehicle, the speedometer value changes through time, and you control that with the accelerator pedal or the brake pedal (negative acceleration).

As training wheels, let's restrict ourselves to "uniform" constant acceleration; unless we make that assumption, we would need Calculus to go further, and for some of you that will occur in lecture.

Making a simple *model* for study purposes, let's consider a velocity that is *steadily increasing or decreasing* and see what consequences that would have on the (position vs. time) measurement. Consider the velocity graph below of an object in free-fall:

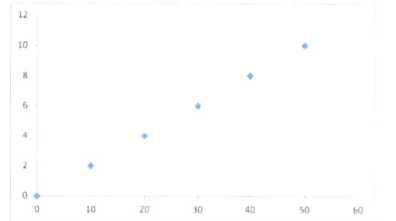 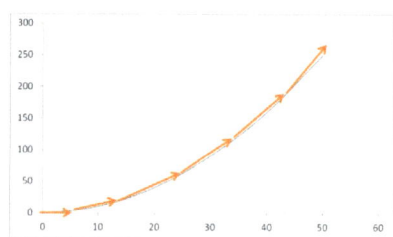

Figure 2-1: Computing the position from the velocity. At each time point on the right graph, the slope there is equal in magnitude to the velocity at that time.

As we learned last week, at each point in time, the velocity graph is the slope of the position graph. Reversing that process, let's plot the value of the slope of vi versus t every 10 s versus time, and call that the position graph, $x(t) = \Sigma\ v_i\ (10\ s)$ where the sum is over time = 0 and the present time.

Start at the first point in the left graph and on the right graph plot a short line with that slope. Follow with the second point in the left graph and on the right graph add a short line with that slope to the previous line. Repeat.

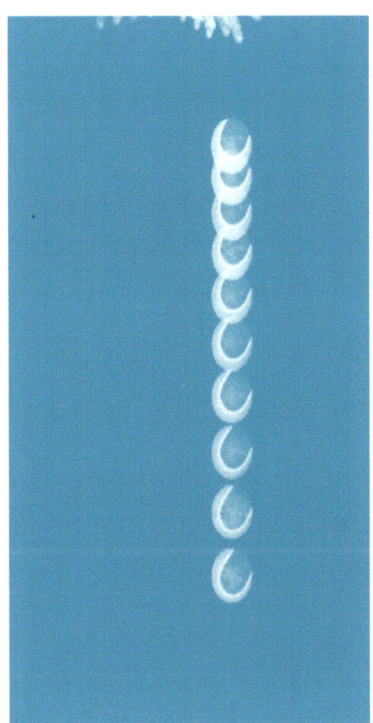

Figure 2-3: Another Representation of an Object in Free Fall: Stroboscopic.

The goal of this experiment is to determine what acceleration graph corresponds to the velocity graph.

To accomplish this task, let's again make a simplifying model and consider "uniform"acceleration" (constant) and see what velocity graph it produces.

Fortunately, for Isaac Newton, there was a readily available source of constant acceleration, the acceleration of objects in free fall near the surface of the Earth. This *special* acceleration, we will call g:

$$a = g = 9.80\ m/s^2$$

<u>Source of Confusion, an Aside:</u>

The concept of Force (a push or pull) is ahead of us. Yet you already have a mental construct (schema) of what this means in everyday life.

You have used the strength of your arms and upper body as a reference point. NOW, you know that if you push briefly as hard as you can on a paper, a book, and your Mother-in-law, you get very different resulting motions.

1. This is a horizontal principle.

<u>Rule 1</u>

This phenomenon is probably stored as, "It is more difficult applying the same force to get *large motion changes* out of a *heavier* object" moving horizontally.

If "you" try moving things in the vertical direction, the rule still works.

2. However, if you drop objects near the surface of the Earth (free fall), the story is different. Ignoring friction, all objects (large and small) fall at the same rate. A detailed study will develop Rule 2. This difference is subtle, difficult to express in words as below, and anti-intuitive!

<u>Rule 2</u>

"Earth" gravity seems to cause *larger motion changes* for heavier objects than for *lighter* objects <u>in free fall</u>, which replaces Rule 1.

As a child, you had less mental hardware, and so *each* of you developed an internal story to cope, and to seek a *simple* answer for this subtle distinction. You buried the distinction between Rules 1 and 2 in your subconscious, and as an adult you may retain a commonly held confusion. Basically, you probably made your World isotropic[14], when in fact motion on Earth is an-isotropic.

3. Now if we simplify one more factor, air resistance, all objects accelerate at the same rate, g, regardless of weight. That means doing the experiment in vacuum, and the technology for which didn't occur until the 1900's.

---

[14] Merriam-Webster Dictionary: Exhibiting properties with the same values along axes in all directions.

Home Experiment: Drop a flat piece of paper and a book at the same time, and the paper lags behind. This reinforces your schema that heavier things fall faster. You can, however, minimize air resistance. Scrunch the paper up into a tight ball (minimizing its contact with air) and the book and paper will accelerate together.

The ultimate reason (this secret will evolve slowly throughout the course) is that heavier objects experience a gravitational force that is proportional to their weight. Try dropping two dice separately, and glued together, and they experience the same acceleration, g.

Fig 2-4 Dice falling independently versus coupled.

Fig 2-5 Skydivers of various masses in free fall at the same rate.

And so we develop Rule 3, a *less* confusing rule than 2:

> Rule 3
>
> All objects (ignoring air resistance) accelerate <u>in free fall</u> at the same rate.

Nevertheless, your subconscious will fight this view and mislead you.

We will be using this "natural" constant acceleration environment (free fall) to complete the investigations about the connection between velocity and acceleration.

Mathematical Description of Motion:

From the scientific method approach, you may have "data" and you want to see if there is some Mathematical Expression that *mostly* describes your data.

But there is far too much Math lying around.

"You" have to make an ***initial guess*** as to what function "fits" your data. That means becoming familiar with function shapes. Then you use that function-fit, and eyeball if the fit is a match to your data. More precisely, there is a *correlation coefficient* which assesses the "goodness" of your fit.

If the correlation coefficient ρ is near zero (0), then you made a bad guess; if the correlation coefficient is near one (1), then you made a good guess.

There also is a root-mean-square error **RMSE**, which gives information about how much the data is scattered about the curve. RSME = 0 for a perfect fit, and RSME is large for either an improper guess of the fitting function, or because the data is polluted by a large source of error and RSME becomes large.

The RSME is a complicated estimate of the *cumulative area error* between the "actual" curve and the "fit" curve.

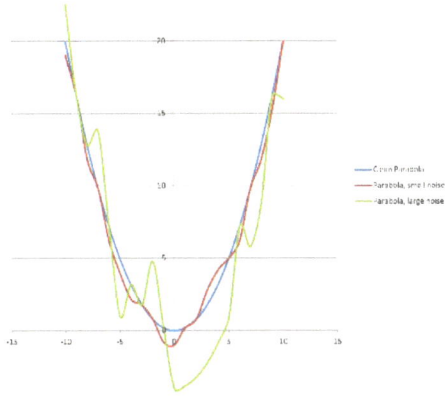

Fig 2-6a Parabola with zero, small and large noise added.

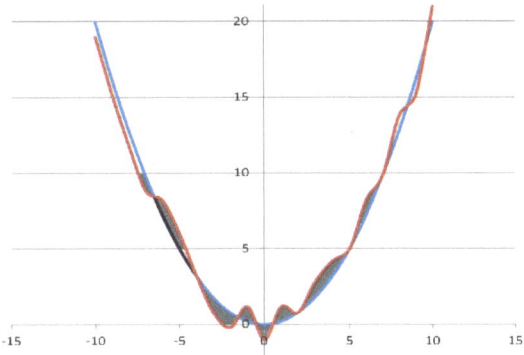

Fig 2-6b RMSE is an estimate of the grey area, counted positive if above or below the data.

A short-list of functions you may wish to guess as a fitting function in the Laboratory is given below:

Constant Function

Linear Function

Parabolic Function

4th Order Polynomial Function

4th Order Polynomial Function

Sine Function

Root Function

Power Function

➢ *Open Capstone Workbook 2 in the Capstone Labs folder on your desktop. Journal.*

Can you always "guess" the correct fitting function?

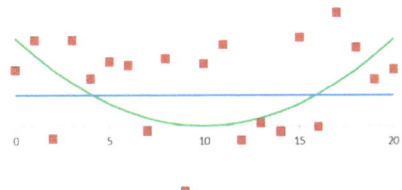

Lab 2

Acceleration

Post-Lab Concept Issues:

Revisit the Pre-Lab questions.

For questions where your thinking has not change, mark NC.

For questions where the experiment has changed your understanding, re-answer the question as you now understand it.

1. The terms distance and displacement are synonymous and may be used interchangeably. Thus the distance an object travels and its displacement are always the same.

2. Velocity is another word for speed. An object's speed and velocity are always the same.

3. Acceleration and speed are the same thing.

4. Acceleration always means that an object is speeding up.

5. Acceleration always occurs in the same direction as an object is moving.

6. If an object has a speed of zero (even instantaneously), it has no acceleration.

Lab 3: Projectile Motion

Name _____

Pre-Lab Concept Issues:

Please respond to the following questions without reading ahead in the Lab book or your Text book. Like free-association, just put down on paper what your gut-level response is:

1. A projected mass initially moves in the direction of firing. Only after some impetus[15] has been used up, can gravity act and the object fall towards the ground.

2. An object that is dropped from a moving carrier does not receive any impetus, and therefore tends to drop straight down. However, air resistance and the speed of the carrier might affect the actual direction of motion.

3. Falling objects possess more gravity than stationary objects, which may possess none at all.

---

[15] Pre-Newtonian *impetus theory* of motion. Put briefly, this theory attributes motion to an impetus that is given to an object initially and is then gradually used up over time.

Lab 3

Projectile Motion

Equipment:

Projectile Launcher
Photo gates
Time of Flight Pad
Computer Interface

Introduction:

The cannon (and before that the catapult) have long been a major tool of warfare.

Major rulers throughout the ages have kept scientists around (one each) to constantly improve the weapon. DE Vinci and Newton were such weapon makers.

Figure 3-1: Cannon in progressive times in History.

The projectile equations are also a major training wheels experiment for Introductory Physics I.

In Lecture, you will (or have already) derived the basic vector equation for any piece of matter in two-dimensions:

$$\vec{r} = \vec{r_0} + \vec{v_0}\,t + \frac{1}{2}\,\vec{a}\,t^2 \quad where\ for\ our\ purposes: \vec{r}(t) = x(t)\hat{i} + y(t)\hat{j}$$

In two-dimensions, these equations are x-y direction- independent (except t, which "drives" the motion) and become.

$$x(t) = x_0 + v_{0x}t + \frac{1}{2}a_x t^2$$

$$y(t) = y_0 + v_{0y}t + \frac{1}{2}a_y t^2$$

where

x_0, y_0 are the positions in x-y land at time t=0

v_{0x}, v_{0y} are the velocities in x-y land at time t=0

$a_x(t)$, $a_y(t)$ are the accelerations in x-y land.

Note: No y-value is in the x-equation, and no x-value is in the y-equation!

These equations are formidable, and suitable for rocketry if one adds a third dimension.

Thus we will make some reasonable assumptions: The projectile is considered as:

A POINT MASS IN FREE FALL

THERE IS NO a_x

which amounts to no appreciable air friction/resistance.

$$a_x = 0, \quad a_y = -g = -9.80 \text{ m/s}^2$$

What follows DEPENDS ENTIRELY on the assumptions above. I.E. if air resistance cannot be neglected, "you", or the Launcher, or the floor is touching the mass you CANNOT use the equations which follow:

$$x(t) = x_0 + v_{0x}t$$

$$v_x = v_{0x}$$

$$y(t) = y_0 + v_{0y}t + \frac{1}{2}(-g)t^2$$

$$v_y(t) = v_{0y} + (-g)t$$

The x_0, y_0 pair are determined by the position vector in x-y space of the projectile at launch (t=0).

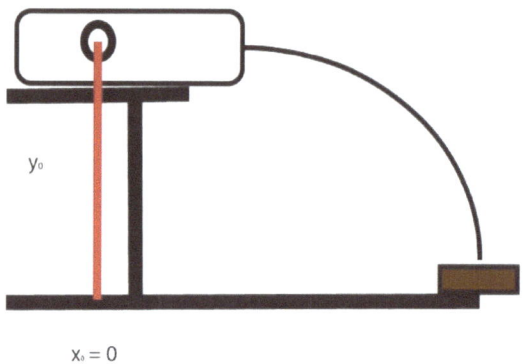

y_0

$x_0 = 0$

Figure 3-2 Establishing the x_0, y_0 initial pair at time = 0.

The v_{0x}, v_{0y} pair are determined, usually, in r-θ coordinates or in x-y components in velocity space. Because the usual *control* the cannoneer has is the elevation angle and launch velocity (amount of explosive).

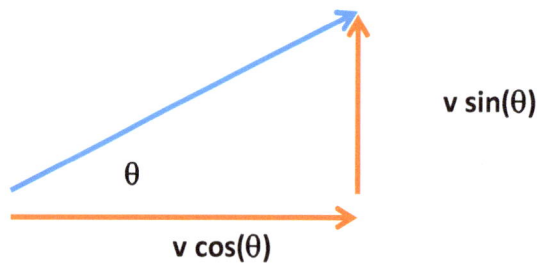

v sin(θ)

θ

v cos(θ)

Figure 3-3: Establishing the v_{0x} and v_{0y} pair geometrically.

Note: These two variables are imbedded in the same x-y coordinate system, but describe different physical concepts (note units), and should be drawn in a <u>separate vector diagram</u>.

What results?

Figure 4-5: Parabolic motion in x-y direction when water drops are in free fall.

➢ *Open Capstone Workbook 3 in the Capstone Labs folder on your desktop and Journal.*

Name _____

Lab 3

Projectile Motion

<u>Post-Lab Concept Issues</u>:

Revisit the Pre-Lab questions.

For questions where your thinking has not change, mark NC.

For questions where the experiment has changed your understanding, re-answer the question as you now understand it.

1. A projected mass initially moves in the direction of firing. Only after some impetus[16] has been used up can gravity act and the object fall towards the ground.

2. An object that is dropped from a moving carrier does not receive any impetus, and therefore tends to drop straight down. However, air resistance and the speed of the carrier might affect the actual direction of motion.

3. Falling objects possess more gravity than stationary objects, which may possess none at all.

---

[16] Pre-Newtonian *impetus theory* of motion. Put briefly, this theory attributes motion to an impetus that is given to an object initially and is then gradually used up over time.

Lab 4: Newton's Three Laws

Name _____

<u>Pre-Lab Concept Issues:</u>

Please respond to the following questions without reading ahead in the Lab book or your Text book. Like free-association, just put down on paper what your gut-level response is:

1. The only "natural" motion is for an object to be at rest.

2. If an object is at rest, no forces are acting on the object.

3. Force is a property of an object. An object contains force and when it runs out of force it stops moving.

4. The motion of an object is always in the direction of the net force applied to the object.

5. Large objects exert a greater force than small objects.

6. A force is needed to keep an object moving with a constant speed.

7. Rocket propulsion is due to exhaust gases pushing on something behind the rocket, like the Earth.

Lab 4

Newton's Three Laws

Equipment:

Motion sensor
Dynamics carts
Friction Accessory
Hover puck
2.0 m Dynamics track
Force sensor
Magnetic end stop
String
Pulley
Hanging mass set
Cart launcher

Introduction:

These explorations will deal with the concept of FORCE.

A preliminary superficial definition might be a "push" or a "pull".

> "In physics, a **force** is any influence that causes an object to undergo a certain change, either concerning its movement, direction, or geometrical construction. In other words, a force can cause an object with mass to change its velocity (which includes to begin moving from a state of rest), i.e., to accelerate, or a flexible object to deform, or both. Force can also be described by intuitive concepts such as a push or a pull. A force has both magnitude and direction, making it a vector quantity. It is measured in the SI unit Newton, and represented by the symbol $\vec{F}$."[17]

---

[17] Wikipedia.

The above notation for force is concise and correct, but the vector notation hides some important details.

- $\vec{F}$ means the **net force**, or the VECTOR sum of all the forces acting on the object

$$\vec{F}_{net} = \sum \vec{F}_i$$

- $\vec{F} = \vec{C}$ means that:

$$\Sigma F_x = C_x \quad \text{AND} \quad \Sigma F_y = C_y \quad \text{AND} \quad \Sigma F_z = C_z$$

Definitions

Inertia: a) Tendency for things to stay put when "at rest", or b) to stay moving at the "same speed and direction".

Mass: Both the amount of matter inside an object (microscopic), and a quantitative measurement of inertia at low speeds. An amendment will be added next Semester.

Newton's Laws

Law 1: If there is NO external force exerted on a piece of matter, then it will stay "at rest" or it will maintain a "constant velocity". Equation matching words:

$$\vec{v} = \vec{c} \qquad \text{a constant vector, possibly } \vec{0} \qquad \text{(1)}$$

➤ **Open Capstone Workbook 4A in the Capstone Labs folder on your desktop and Journal.**

Law 2: If there **IS** a non-zero net force exerted on a piece of matter, then the acceleration it attains is proportional to the strength of the net force, and inversely proportional to its mass.

$$\vec{a} = \vec{F} / m \qquad \text{(2)}$$

Validity of Newton's Second Law

Newton's Second Law as written above is valid when measurements are made in an "inertial" coordinate system, i.e., a coordinate system at rest or moving with constant velocity.

If, however, the object and the observer are accelerating *relative* to each other, the Law as stated above is not the full story. Enter: the "fictitious force".

When measurements are made in an accelerating coordinate system of an object not embedded in *that* system, the true motion of a mass in its own coordinate system is distorted to the observer in the accelerating coordinate system.

Case A: Acceleration parallel to the line of motion, speeding up

"Consider an **accelerating** transit car with a weight hanging inside it, suspended from the ceiling, as shown.

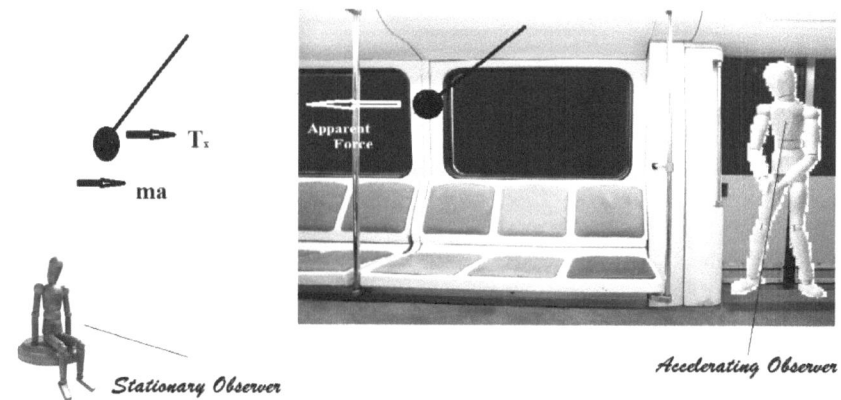

Figure 4.6 Stationary and Non-stationary observers interpretation.

A. We know that **F** = m**a** is true for an **observer on the ground**. That observer "sees" the forces we have sketched above on the left. The suspended weight does not hang straight down because the tension provides a horizontal component of force to give it an acceleration (and a vertical component to counter its weight). This "observer on the ground" may be referred to as an **inertial observer** or an observer in an **inertial reference frame** and Newton's Second Law of Motion is valid in her frame of reference.

B. What will be seen by an **observer inside** this accelerating railroad car?

The hanging weight does not hang straight down. It is still suspended by the cord at an angle θ as shown. And the weight is **at rest** with respect to this "onboard observer"!

If it is at rest, how can it hang suspended like that? We believe so strongly in the Newton's Second Law so that we "invent" another force. We call this a "fictitious force", at least in the view of the stationary observer.

Such fictitious forces – or non-inertial forces – occur in **accelerating reference frames.** An accelerating reference frame is a **non-inertial reference frame.**

That simply means that the Law of Inertia is **not true** in that frame unless these fictitious forces, or inertial forces are introduced."

Case B: The observer is accelerating in a perpendicular direction to the line of motion, executing curved-line motion

Suppose you have a "loose" (meaning not bound by friction) package in the rear of your auto, and you (with seat belt on attaching you to the auto and its motion) steer the car left. This is an acceleration because the direction of velocity has changed.

In an inertial coordinate system, the package travels in a straight line, due to Newton's First Law. However, to you as an accelerating driver, it "appears" as though the package accelerates (and that requires a force?) toward the right side of your vehicle. The apparent force is "fictitious".

Left: a crate near the driver's side back wheel well will travel in a straight line (Newton's 1st) and remain near that wheel well.

Right: if the truck on the right carries a similar crate (and the truck bed is slippery), then it will also travel in a straight line (Newton's 1st Law, no net force) in "our" coordinate system, but the driver of the right truck will see the crate slide from one wheel well to the other in the rear view mirror compared to "their" coordinate system.

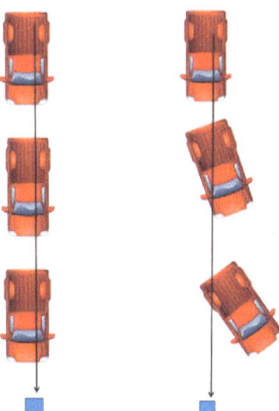

Figure 5-2 "Loose" objects travel in a straight line, according to Newton's First Law, but appear to move laterally to the accelerating observer

Follow the straight lines on both columns and deduce, in each case, where a frictionless object would go relative to each driver.

Open Capstone Workbook 4B in the Capstone Labs folder on your desktop and Journal.

Law 3:   There ARE no isolated forces. Forces come in pairs from at least two pieces of matter. If body A exerts a Force$_A$ on body B, then body B will exert the same magnitude of force (but in the opposite direction) Force$_B$ on A. Always!

$$\vec{F}_{A\,on\,B} = -\vec{F}_{B\,on\,A} \qquad \textbf{(3)}$$

Figure 5-3: An illustration of Newton's third law in which two skaters push against each other. The first skater on the right exerts a normal force N$_{12}$ on the second skater with her hand directed towards the left, and the second skater exerts a normal force N$_{21}$ on the first skater directed towards the right with her back. The magnitude of both forces are equal, but they have opposite directions, as dictated by Newton's third law.

Validity of Newton's Third Law

Newton's Third Law is valid **in all coordinate systems**, including accelerating observers. However, **one** of the Third Law pair may be a fictitious force if the observer is accelerating.

Again, consider yourself as the driver of an auto. When your auto is **at rest**, your seat back applies a small force on you if it is tilted. However, if you make your seat back completely vertical (uncomfortable) the seat back will apply no force on you. Like standing with your back to a flat wall at home.

If you now punch the accelerator, both you (the observer) and the auto accelerate. In order for you to accelerate, there must be a net horizontal force pushing you forward (Newton's Second Law), and that is now applied by your seat back. If you imagine putting your hand between the seat back and your back, you will find a force on BOTH sides of your hand and these are equal and opposite. Newton's Third Law.

If this were not true, either the seat back would accelerate more than you and potentially break you; or, you would accelerate less than the seat back and you would potentially break your seat.

➤ *Open Capstone Workbook 4C in the Capstone Labs folder on your desktop and Journal.*

Lab 4

Newton's Three Laws

Post-Lab Concept Issues:

Revisit the Pre-Lab questions.

For questions where your thinking has not change, mark NC.

For questions where the experiment has changed your understanding, re-answer the question as you now understand it.

1. The only "natural" motion is for an object to be at rest.

2. If an object is at rest, no forces are acting on the object.

3. Force is a property of an object. An object contains force and when it runs out of force it stops moving.

4. The motion of an object is always in the direction of the net force applied to the object.

5. Large objects exert a greater force than small objects.

6. A force is needed to keep an object moving with a constant speed.

7. Rocket propulsion is due to exhaust gases pushing on something behind the rocket like Earth.

Lab 5: Adding Force Vectors & Static Equilibrium

Name _____

Pre-Lab Concept Issues:

Please respond to the following questions without reading ahead in the Lab book or your Text book. Like free-association, just put down on paper what your gut-level response is:

1. **Adding Vector** quantities is the same as any other algebraic addition.

2. If several forces are acting on the same mass, there are no methods for determing the "effective" equivalent force.

3. Define equilibrium in words and equations.

Lab 5: Adding Force Vectors & Static Equilibrium

Equipment

Force Table
Pulleys
Metal rings with string
Center pin
Mass hangers
Mass set
Level
Pasco computer interface
Force Sensor
QtiPlot or Origin Software (will plot vectors from XY XY pairs)

Notes on procedures:

- Measure all masses to the tenth of a gram.
- Measure all angles to the nearest tenth of a degree.
- Always consider the total mass that is relevant, in this and other labs. For instance, in this lab you will place given masses on a mass hanger and then hang the mass hanger from a metal ring. The total mass is the deposited mass + mass of hanger . You are responsible for recording and calculating with the *actual* mass. Do not waste time trying to make the total mass come out even.
- Your setup will include a set of metal rings tied together with strings. You do not need to untie the strings: strings that are not used in the procedures can rest on top of the force table. Be sure, however, that the unused strings do not interfere with the hanging masses.
- When hanging more than one mass, be sure the central peg is in place so that your partially built system doesn't accelerate. Make sure to align the pulleys so that the strings are perpendicular to the edge of the table when the center ring is centered on the center pin.

Introduction:

In this laboratory we will investigate the vector nature of forces. We will also investigate what happens when two or more forces are applied to the same object. That is, we will determine the *net force* exerted when more than one force is applied. The net force is that single force that has the same effect as two or more forces applied at the same time, sometimes called the **Resultant**.

Weight and mass

The forces we will use will be the weights of various masses. The weight of any mass is given by:

$$W = mg$$

$$(g = 9.80 m/s^2)$$

where W is the weight in Newton's, m is the mass in kilograms and g is the acceleration of gravity. The above expression follows from Newton's Law of Gravitation and Newton's second Law of Motion. *This form is only valid near the surface of the Earth.*

Vectors

It is an experimental fact (which we will verify) that forces ***cannot*** be adequately described by a single signed number if the other forces exist in multiple directions. To describe a 2D ***force***, we must specify two quantities: a magnitude and a direction. Physical quantities that have direction as well as magnitude are called **vectors**. Other vector quantities are: displacement, velocity, acceleration, and momentum. Physical quantities that can be described by a magnitude only are called **scalars**. Examples of scalars are mass, length, time, temperature and energy.

Graphically, we can represent force vectors (or any other vector quantity) by arrows, as in the diagrams below. The length of the arrow indicates the strength of the forces attached to a point-mass in space, and the direction represents the true direction.

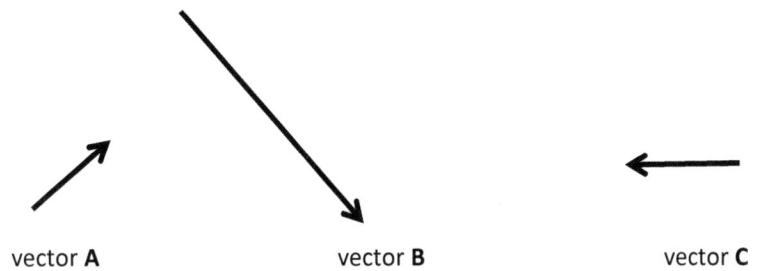

vector **A**   vector **B**   vector **C**

If the vectors lie in one or two dimensions only, we can use the **graphical method** for adding vectors. Recall that vectors are "sliders". We first set up an x-y coordinate system on paper and designate a convenient scale for the coordinate axes. We then draw to scale the vectors to be added, placing the tail of the first vector at the mass, the tail of the second vector at the head of the first, the tail of the third at the head of the second, and so on:

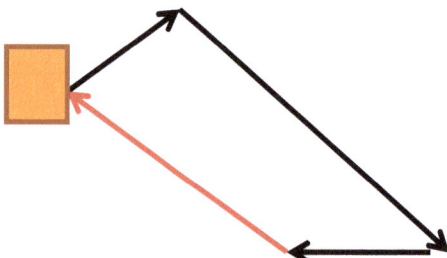

Suppose vectors A, B, C are acting on a mass. We add them as above.

Now what vector **D** is need to be added to A, B, C so that the sum of A, B, C, **and** D add to zero? The red one shown!

This will be our process: we will add N forces to a ring (through hanging weights) and predict and what additional force will bring the ring into equilibrium; then we will experimentally measure the N+1th force required.

Static Equilibrium:

From Newton's Second Law,

$$0 = \Sigma \vec{F} = m\vec{a} \implies \vec{a} = 0 \ \ for \ any \ non-zero \ mass \quad \textbf{(1)}$$

In three spatial dimensions we write

$$\Sigma \vec{F} = \vec{0} \qquad which \ is \ actually \ the \ 3D \ number \ (0,0,0) \quad \textbf{(2)}$$

The vector notation is concise, but it can be deceiving. In component form the above equation means

$$\Sigma F_x = 0 \ \ \& \ \ \Sigma F_y = 0 \ \ \& \ \ \Sigma F_z = 0 \qquad \textbf{(3)}$$

A more practical (and precise) way to add vectors is the **analytical method**, which can be used for vectors in any number of dimensions. This consists of three steps:

1. Break each vector to be added into its components along the Principle Axes.
2. Add the components in each axis direction independently.
3. Find the resultant from the sum of the components.

$$\vec{R} = R_x\,\hat{i} + R_y\,\hat{j} + R_z\,\hat{k}$$

1. Breaking a vector into its components

If we place the vector **A** at the origin of a coordinate system, then **A** forms a right triangle with the x- and y-axes (see Diagram 2). The **components** of **A** are those special vectors that are parallel with the coordinate axes and whose vector sum is **A**. A vector in two dimensions will have two components; a vector in three dimensions will have three components, and so on.

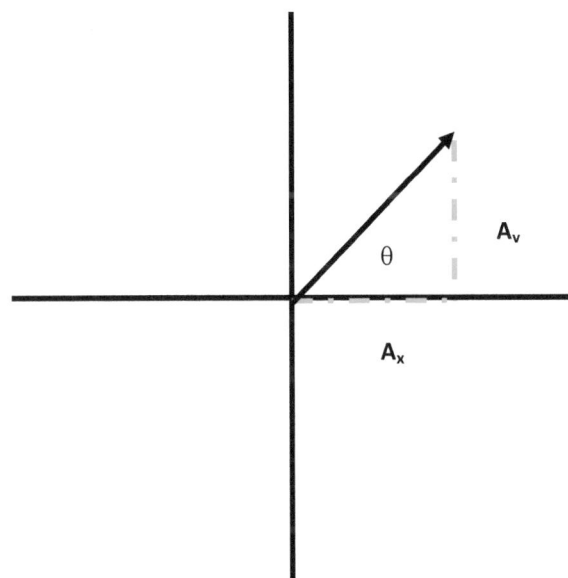

In the diagram, the dashed line A$_x$ is the x-component of **A**; the dashed line A$_y$ is the y-component of **A**. From trigonometry and from the Pythagorean Theorem we know that:

(1) $A_x = A\cos\theta$ and $A_y = A\sin\theta$

(2) $A^2 = A_x^2 + A_y^2$ and $\tan\theta = \dfrac{A_y}{A_x}$

Thus if we know the magnitude and direction of **A**, we can find its components (A_x and A_y);

conversely, if we know the components of **A**, we can find its magnitude and direction (A and θ).

The first step in adding vectors, then, is to break each one into x- and y-components using equations 3 above. Note: A vector need not be located at the origin for us to do this. We simply find the *projection*, or shadow, of the vector on the coordinate axes.

2. Adding x- and y-components

Look at the Figure below, where the vectors **A** and **B** are added graphically. The resultant is labeled **R**. By examining the diagram (which is for two vectors), you can extrapolate that:

- The x-component of the net force is the sum of the x-components of **A** and **B** and **C**:

$$\Sigma F_x = A_x + B_x$$

- The y-component of the net force is the sum of the y-components of **A** and **B**:

$$\Sigma F_y = A_y + B_y$$

This method works for any number of vectors.

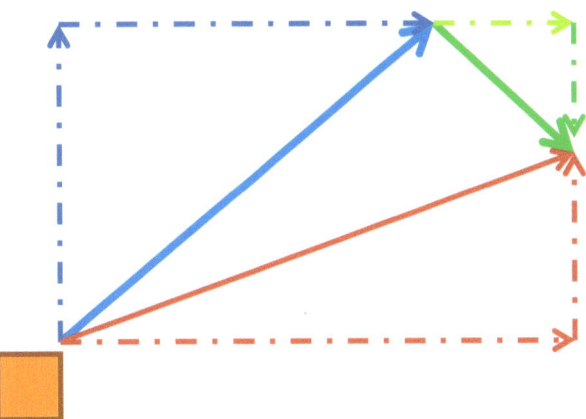

In the diagram above, vector A has its tail attached to the mass. Vector B is added tail to A. The vector sum A + B is the red vector. The red dashed lines are the x- and y-components of A + B.

The blue dashed lines are the x- and y-components of A; the green dashed lines are the x- and y-components of B.

Note that the sum of the blue and green components x-component is the red x-component. Similarly for the y-direction

Experimental apparatus: The force table is a circular platform with a degree scale around its circumference. Pulleys clamped to the table allow us to hang weights at various angles. The weights are connected by means of strings to a metal ring at the center of the table; a metal pin keeps the ring in place when the forces are NOT in equilibrium. When the hanging weights are in equilibrium, however, the pin can be removed since no net force acts on the ring, and the ring will remain at rest.

Activity I: Addition of Two Vector Forces when the ring is in static equilibrium

1. Set up the force table with one pulley and with the center pin inserted into the center of the platform. Adjust the feet so that the table is level with the ground.

2. Place a pulley at an arbitrary angle and drape a string from the ring over it. Hang a mass hanger from the string and place a mass of 200-grams (on the 50-gram hanger). Observe the effect on the center ring.

3. Calculate and record the total hanging mass on pulley 1. Record the angle θ, and the weight of **F** in row 1.

Vector Force	Hanging mass including weight hanger **m(kg)**	**F = mg (N)**	θ
1			
2	NA		

4. Attach the Force Sensor horizontally to the ring, and move it in angle and strength until the ring no longer touches the post. Record the Force applied and the angle of the Force sensor in row 2.

Analysis

a. Open the EXCEL file "Vector Data" in the Lab 5 Capstone Folder.

b. Enter the pair of data from the right two columns into the places to the right of each the Vector numbers in the pink data cells.

c. EXCEL will calculate the 2-D vectors and components.

d. Highlight and copy all the numbers in columns D – G, red.

e. Open QiData on the desk top.

f. Paste your data into the QtiPlot data table.

g. Highlight the data by clicking on the column headers.

h. Under the Plot pull-down menu choose Vectors XYXY

i. Your 2-D graph of the vectors will be drawn. Fill Screen. Print for your report.

Measure Vectors 1 - 2 and the distance between the tail of Vector 1 and the head of

   Vector 2 on the graph.

See an example below with three hanging masses, and a 4th force you apply to bring the system to equilibrium. The gap between the tail of vector 1 and the head of vector 4 is the experimental error.

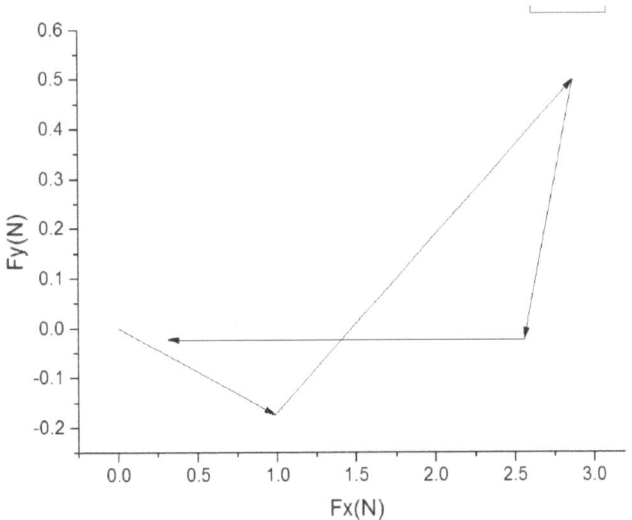

Vector 1 length   _____ (units)

Vector 2 length   _____ (units)

Vector gap length _____ (units)

The vector gap is the error and the Vector 1 length is the reference. Calculate your percent error.

% error = (gap length/length-of-Vector-1) x 100% = _____ %

Close the application without saving.

Activity 2: Addition of Three Vector Forces when the ring is in static equilibrium

1. Place another weight hanger with 40 g on the table at some arbitrary angle less than 180^0 from the first weight hanger. Transfer row 1 from the table above.

2. Calculate and record the total hanging mass on pulley 2. Record the angle θ and **F** in row 2 for the current

hanging mass.

Vector Force	Hanging mass including weight hanger **m(kg)**	**F = mg (N)**	θ
1			
2			
3	NA		

3. Attach the Force Sensor to the ring, and move it in angle and strength until the ring no longer touches the post. Record the Force applied and the angle of the Force sensor in row 3.

Analysis

a. Open the EXCEL file "Vector Data" in the Lab 5 Capstone Folder.

b. Enter the pair of data from the right two columns into the places to the right of each the Vector numbers in the pink data cells.

c. EXCEL will calculate the 2-D vectors and components.

d. Highlight and copy all the numbers in columns D – G, red.

e. Open QiData on the desk top.

f. Paste your data into the QtiPlot data table.

g. Highlight the data by clicking on the column headers.

h. Under the Plot pull-down menu choose Vectors XYXY

i. Your 2-D graph of the vectors will be drawn. Fill Screen. Print for your report.

Measure Vectors 1 - 3 and the distance between the tail of Vector 1 and the head of

Vector 3 on the graph.

Vector 1 length   _____ (units)

Vector 2 length   _____ (units)

Vector 3 length   _____ (units)

Vector gap length _____ (units)

The vector gap length is the error and the Vector 1 length is the reference. Calculate your percent error.

% error = (gap length/length-of-Vector-1) x 100% = _____ %

Close the application.

Activity 3: Addition of Four Vector Forces when the ring is in static equilibrium

1. Place 130 grams on a Hanger 3 at another arbitrary angle between 180^0 and 270^0.

2. Bring forward the first two rows in the Table above. Record your new hanging mass and calculate the F and θ in row 3.

Vector Force	Hanging mass including weight hanger m(kg)	F = mg (N)	θ
1			
2			
3			
4	NA		

3. Attach the Force Sensor to the ring, and move it in angle and strength until the ring no longer touches the post. Record the Force applied and the angle of the Force sensor in row 4.

Analysis

a. Open the EXCEL file "Vector Data" in the Lab 5 Capstone Folder.

b. Enter the pair of data from the right two columns into the places to the right of each the Vector numbers in the pink data cells.

c. EXCEL will calculate the 2-D vectors and components.

d. Highlight and copy all the numbers in columns D – G, red.

e. Open QiData on the desk top.

f. Paste your data into the QtiPlot data table.

g. Highlight the data by clicking on the column headers.

h. Under the Plot pull-down menu choose Vectors XYXY

i. Your 2-D graph of the vectors will be drawn. Fill Screen. Print for your report.

Measure Vectors 1 - 4 and the gap distance between the tail of Vector 1 and the head of

Vector 4 on the graph.

Vector 1 length   _____ (units)

Vector 2 length   _____ (units)

Vector 3 length   _____ (units)

Vector 4 length   _____ (units)

Vector gap length _____ (units)

The vector gap length is the error and the Vector 1 length is the reference. Calculate your percent error.

% error = (gap length/length-of-Vector-1) x 100% = _____ %

Delete all the data and the graph in QiData or close the application.

Conclusions:

1. In each force addition setup, was the gap length (the error) small compared to the Vector 1 length?

2. Were your percent errors acceptable?

3. Do you think you proved Eqs. (2) and (3) above?

Name _____

Lab 5

Adding Force Vectors & Static Equilibrium

Post-Lab Concept Issues:

Revisit the Pre-Lab questions.

For questions where your thinking has not change, mark NC.

For questions where the experiment has changed your understanding, re-answer the question as you now understand it.

1. **Adding Vector** quantities is the same as any other algebraic addition.

2. If several forces are acting on the same mass, there are no methods for determing the "effective" equivalent force.

3. Define equilibrium in words and equations.

Lab 6: Conservation of Energy

Name _____

Pre-Lab Concept Issues:

Please respond to the following questions without reading ahead in the Lab book or your Text book. Like free-association, just put down on paper what your gut-level response is:

1. Is there a difference between "force" and "energy"?

2. Can an object at rest have a non-zero energy?

3. Is energy used up by objects? If yes, explain how.

Lab 6: Conservation of Energy

Equipment:

2.0 m Dynamics track
Dynamics cart with Force sensor attached
Pulley
Heavy weight String
Spring
Hanging Weights

Introduction:

There are several kinds of interactions that CANNOT be described by the previous techniques.

Situations:

1. We get a <u>distribution</u> of solutions rather than a <u>unique</u> solution.

Collision of "real extended" rather than "point" objects

2. Roughness – because the EXACT details of surface atoms matter, even doing your very best to shoot the cue ball the same every time, the result is slightly different each time.

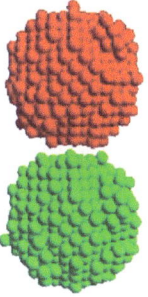

Figure 7-1 Atomic Details[18]

---

[18] Hiroto Kuninaka and Hisao Hayakawa, <u>Phys. Rev. E **79**, 031309 (2009)</u>, Published March 30, 2009

3. Deformation

Objects that deform are usually warmer afterwards. Maybe a little, maybe a lot. To try this out, place a rubber band on your lip (your most sensitive temperature organ) and stretch it.

What do you feel?

Deformation is often difficult to model with any precision for finite objects.

4. Friction

When matter in any one of its states (solid, liquid, gas) rubs against other matter, heat is also produced. To try this out, rub your palms together briskly (like on cold days). What do you observe?

Friction cannot be quantified very accurately. What exactly happens depends <u>exactly</u> and <u>microscopically</u> (atomic level) upon the surface roughness of the two materials.

Below: Two microscopically surfaces are shown: the right one is stationary and the left one is moving to the upwards (blue arrow).

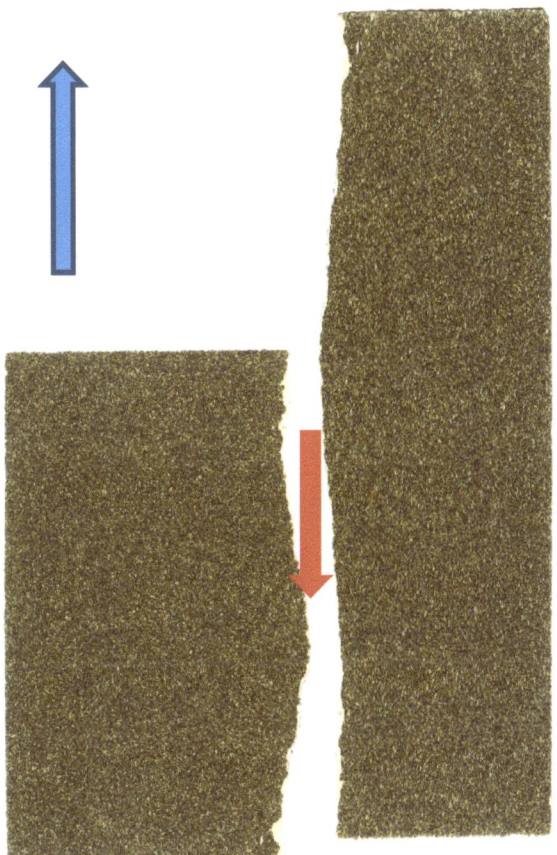

Notice that the bulge on the left one will collide with a bulge on the right one (red arrow). The right one will exert a downward force on the left one, and slow it down.

5. We <u>DON'T KNOW the mathematical form</u> of, for instance, the Nuclear Force nor with any precision the Frictional Force, so Newton's Laws are useless!

$F_g = G\, m_1 m_2 / r^2$

$F_e = k\, q_1\, q_2 / r^2$

$F_n = ?????$

Crude model: $f = \mu N$ good to 10% or so

6. Suppose we are dealing with a <u>complicated system</u> of a <u>large number of moving, possibly colliding,</u> masses and the forces are complex and ever-changing. To measure this system in practice is not possible.

Probing Matter with Particles

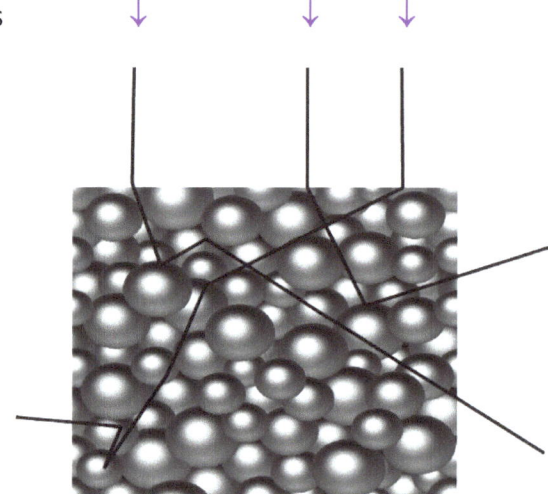

7. Suppose we <u>only</u> KNOW F(x)

If we knew x(t) we could form F(x[t]), but then we already know x(t), so no need!

8. Suppose we only want to KNOW properties of a mass (say velocity) at certain x's.

A

B

C

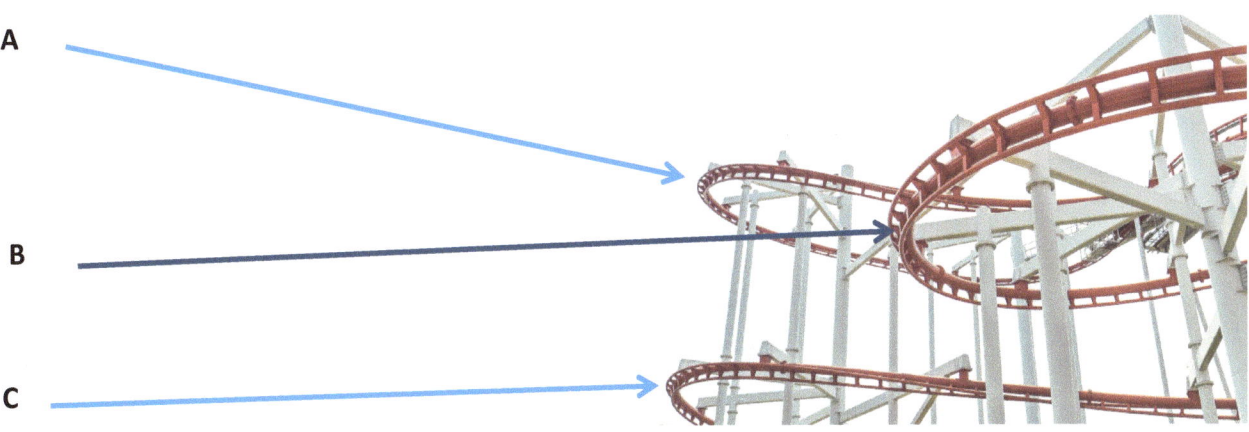

How do we deal with these all too real situations? Solution: "Black Box Physics"

Isaac Newton (1642-1727) was a great mathematician and a great physicist, and probably the most incisive thinker ever known. He chiefly established that natural phenomena generally follow determinate mathematical laws consistent laws of motion, of gravity and of other phenomena. He produced his 'black box' theory of science as explaining only *how* things happen but not necessarily *why* things happen (Newton did not have access to Atomic level Physics then).

BLACK BOX

Stimulus/Input   Response/Output

The Black Box Process often allows us to determine what <u>roughly will come out</u> of the "Black" Box without knowing what is inside. Often, it tells us what <u>cannot come out,</u> and <u>sometimes</u> allow us to <u>deduce what is in the black box.</u>

The First of these "Tools":   WORK

WORK, at this point, is only a reformulation of Kinematics equations and Newton's Second Law:

$$2a\Delta x = v^2 - v_0^2 \qquad\qquad F = ma$$

Steps leading to the definition of Work:

$a = F/m$

$2(F/m)\Delta x = v^2 - v_0^2$

Multiplying by m/2

$$F \Delta x = \Delta W = \tfrac{1}{2}\, m(v^2 - v_0^2) = \Delta K$$

So, a force acting through a displacement causes a change in kinetic energy.

Until now, we have been dealing with F(t).

We will return to that view in other labs.

The Second of these "Tools":   ENERGY

Definition of Conservation:

The term "conservation" has changed meanings over the last century, mainly due to overpopulation and finite resources. Modern: conservation is "hoarding" or "saving", because resources have become scarce.

During the Last Century, conservation meant "the amount remains the same".

Definition of Energy

ENERGY = ABILITY TO CAUSE CHANGES IN MOTION; THE 'PRICE WE PAY' FOR CAUSING CHANGES IN MOTION (can be translated into $)

There are 6 kinds of energy (so far); the total energy at any time is the sum of these energy types:

$$E_{tot} = \Sigma\, (K + U + Q + CE/EE\ (V) + N_E + RE)$$

- ➢ K = energy of directed motion = $\tfrac{1}{2}\, mv^2$ for one mass
- ➢ U = energy of position = - W(to change position)

    Special case, gravity and height near the Earth's surface:

U_g = Weight•height = mgh

➢ Q = energy of (random) internal K of multiple internal masses.

Brownian motion: the random movement of molecules due to their temperature (random direction kinetic energy). Ink particles from the bottom migrate upwards, colliding into water molecules. The higher the temperature, the faster the diffusion.

➢ qV = electric analogs of above**
➢ N_e = energy stored in attraction of nuclei**
➢ RE = Rest Energy -- creation of mass from energy = m_0c^2**

** Next semester

Energy can be thought of as an EXCEL spreadsheet, whose columns represents the amount of each energy form, and then the quantitative sum of those columns.

I.E., a mind picture of a woo-woo fluid that is "conserved", flows between masses freely, and has several personalities.

Energy Conversion During Events:

To the left, water has gravitational potential energy at the top of the sluice box. As it runs down the sluice, potential transforms into translational kinetic energy. Running water strikes the paddle wheel and converts to rotational kinetic energy. That moves into the mill to perform work on grain (grinding)

Activity 1: Suppose we only have two kinds of energy in play: Kinetic and Gravitational Energy

$$E_A = E_B = E_C = E_C = \ldots$$

where A, B, C, D … are various times or spatial locations throughout the motion. This energy may change its form freely.

For this simple ideal system, we can write Mechanical Energy Conservation of the roller coaster car as:

$$\tfrac{1}{2}\,mv_A^2 + mgh_A = \tfrac{1}{2}\,mv_B^2 + mgh_B = \tfrac{1}{2}\,mv_C^2 + mgh_C \;\ldots$$

For non-ideal systems, friction exists, so we write:

$$\tfrac{1}{2}\,mv_A^2 + mgh_A + U_{int}(A) = \tfrac{1}{2}\,mv_B^2 + mgh_B + U_{int}(B) = \tfrac{1}{2}\,mv_C^2 + mgh_C + U_{int}(C) \;\ldots$$

U_{int} is the energy associated with $f = \mu N$, $U_{int} = -W = -\mu N \Delta x$.

The traditional experiment using the above equations involves a cart on a track, a pulley, and a hanging mass.

People often do not understand what the hanging mass is doing (providing a constant force of tension of various strengths), and often confuse the hanging mass and the mass of the cart. BEWARE!

Nevertheless.

$$\tfrac{1}{2}\,mv^2 \Rightarrow T \approx mg\Delta y \approx mg\Delta x$$

So your velocity should behave as $v \approx \sqrt{\Delta x}$ and your kinetic energy should be linear in Δx

> ➤ **Open Experiment 6A in the "Capstone Experiments" folder on the desk top and Journal.**

Activity 2: Suppose we only have two kinds of energy in play: Kinetic and Spring Energy

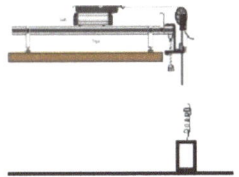

This is a nice example of a non-constant force:

$$F_{spring} = -\,k\Delta x$$

Until now, a non-constant force could not be used in our contant-acceleration equations of motion. However, with energy conservation this varying force can result in computation of motion!

In this case, $U_{spring} = \tfrac{1}{2}\,k\,\Delta x^2$. If we convert spring energy to kinetic energy

$$\tfrac{1}{2}\,mv^2 \Rightarrow \tfrac{1}{2}\,k\,\Delta x^2$$

So your velocity vs. Δx should be a straight line, and your kinetic and potential energies parabolas in Δx.

The total energy then becomes

$$E = \tfrac{1}{2}\,mv^2 + \tfrac{1}{2}\,k\Delta x^2$$

We can test this form of energy conservation using the motion sensor to measure both Δx and v.

> ➤ **Open Experiment 6B in the "Capstone Experiments" folder on the desk top and Journal.**

Name _____

Lab 6

Conservation of Energy

<u>Post-Lab Concept Issues</u>:

Please respond to the following questions again, this time hopefully with more detail. You may also discover that you have changed your mind on an issue: Note that below:

1. Is there a difference between "force" and "energy"?

2. Can an object at rest have a non-zero energy?

3. Is energy used up by objects? If yes, explain how.

Lab 7: Impulse: How External Contact Forces Are *Really* Exerted

Name _____

<u>Pre-Lab Concept Issues:</u>

Please respond to the following questions without reading ahead in the Lab book or your Text book. Like free-association, just put down on paper what your gut-level response is:

1. Stopping an object depends on its velocity.

2. According to the momentum equation, it is harder to stop an object with a bigger mass.

3. It is important that the velocity of the object should be low to stop it.

Lab 7: Impulse: How External Contact Forces Are *Really* Exerted

Equipment:

2.0 m Dynamics Track
Motion Detector, to side of track
Dynamics Cart with Force Sensor
Cart Launcher
End Bracket
Soft and Strong springs
Clay bumper
Two motion carts with mounted force sensors
Rubber bumpers for force sensors
String
Pulley
Mass hanger and masses

Introduction:

We found in a previous lab that a *net Force applied over a distance interval* causes a *change in Kinetic Energy*.

In this Lab, we will find that a *net Force applied over a time interval* causes a *change in Momentum*.

Impulse

Let's start with the relationship between Force and Momentum.

Defining momentum (provisionally a useful quantity)

$$p = mv$$

and our long term friend, Newton's Second Law

$$F_{net} = ma \qquad \text{where } a = \Delta v/\Delta t$$

Then

$$F_{net} = ma = m\, \Delta v/\Delta t$$

Multiplying by Δt

$$F_{net}\,\Delta t = mv_f - mv_i$$

If the mass is constant (you don't throw up or get soaking wet)

$$F_{net}\,\Delta t = \Delta p = mv_2 - mv_1 = p_2 - p_1 \qquad (1)$$

So, a net Force acting over time produces a change in momentum.

In addition, you should be able to show that

$$K = \tfrac{1}{2}\,m\,v^2 = p^2/(2m) \qquad (2)$$

Let's Look At One Ideal Scenario -- A moving object runs into an everyday "solid" object and rebounds.

Depending upon the stiffness of the material ("solid wall", spring, bed mattress, sheet of paper, etc.), the object will necessarily involve a Force applied over a time interval Δt. The "solid" object will deform (see previous Lab).

Given that an object with mass m and velocity v must be collide with the material,

$$F_{net}\,\Delta t = \Delta p = p_2 - p_1 = mv_2 - mv_1$$

Thus, to rebound **this mass**, a large force can be applied over a short time, or a small force can be applied over a longer time. If the mass has a "fracture limit force", then to rebound from the solid to remain intact, we need the maximum force to be less than this limit, and we must increase the time it is exerted.

CUSHION

Fig 7-1 How different forms of impulse matter.

From Eq. (1), we define "Impulse"

$$\text{Impulse} = \vec{J} = \Sigma \vec{F}\Delta t = \vec{p}_2 - \vec{p}_1 \qquad\qquad (3)$$

The sum of $F_i \Delta t_i$ IS the area under the **F** vs. **t** curve.

Suppose you throw a tennis ball at a "solid" wall. If there is no energy loss to other forms (such as heat or sound during the deformation), then $|v_{in}| = |v_{out}|$ and

$$J = m([v_{out}-(v_{in})] = m[-u - (+u)] = -2mu = F_{net}\Delta t$$

And the wall applies a net force $F_{net} = -2mu/\Delta t$.

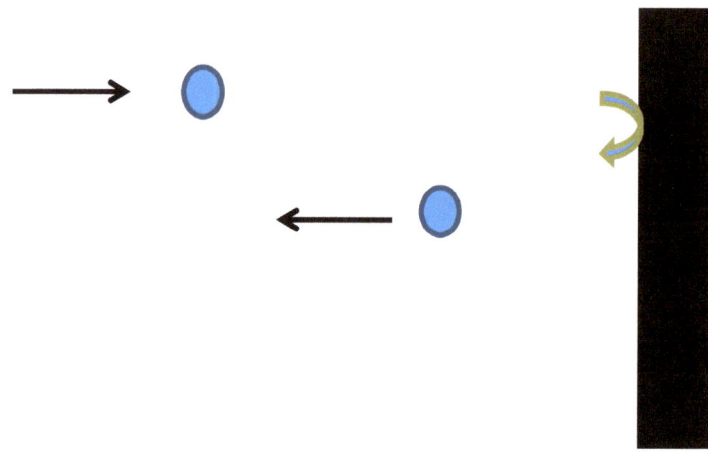

Fig 7-2 Collision with an immovable object

We call this collision <u>perfectly elastic</u>. Also, as a consequence of no other forms of energy in play, kinetic energy is conserved:

$$K_i = \tfrac{1}{2} mu^2 = K_f = \tfrac{1}{2} m(-u)^2$$

A *realistic* Force vs. Time graph is shown:

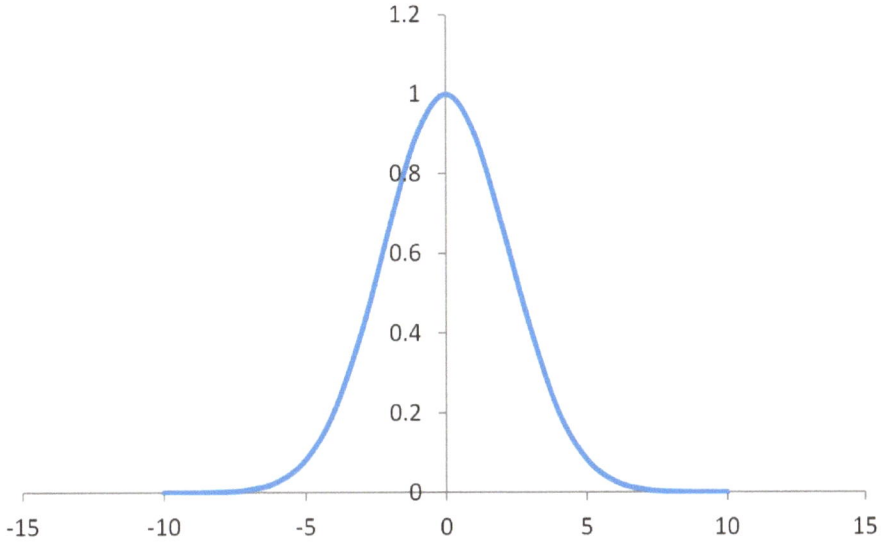

Fig 7-3 Force versus time: Impulse.

A. ***Exploring collisions with various kinds of bumpers against an immovable wall, and the external force they apply***.

   The cause of these rebounds can range from strong (stiff spring), to medium (weak spring), to completely inelastic (the incoming object sticks to the wall).

   ➢ ***Open Capstone Workbook 7A in the Capstone Labs folder on your desktop and Journal.***

B. ***Suppose there are <u>no external forces</u> and there is a collision of two masses. Then there are only internal force pairs, equal and opposite. The contact times for both masses must be equal (why?), the forces must be equal (Newton's Third Law), and so the Impulse graphs (for each force applied to the other mass) should be ±images of each other.***

   ➢ ***Open Capstone Workbook 7B in the Capstone Labs folder on your desktop and Journal.***

C. *Suppose there is a net-Force applied from the left to adjacent objects, which mutually accelerate and do not separate. Does Newton's Third Law STILL apply?*

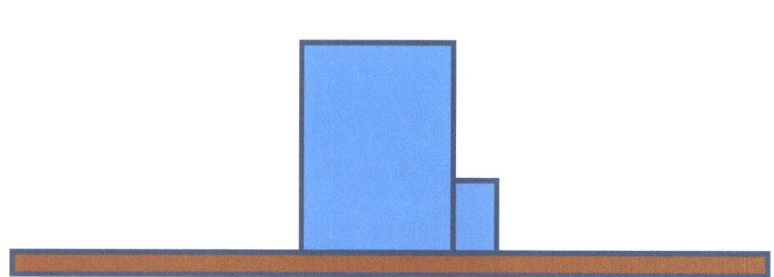

Fig 7-4 Acceleration of coupled masses.

The free body diagrams for the master diagram above are:

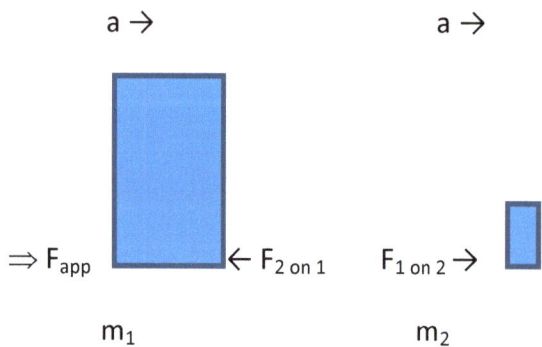

Newton's Second Law Equations for <u>each</u> FBD are:

$$m_1 a = F_{app} - F_{2\ on\ 1} \qquad\qquad m_2\ a = F_{1\ on\ 2}\ (4)$$

Note: because the SIGNS of these internal forces have been established on the FBD's, both $F_{1\ on\ 2}$ and

$F_{2\ on\ 1}$ are positive!

However, if we consider m_1 and m_2 as a single unit, the relevant form of Newton's Second Law is

$$(m_1 + m_2)a = F_{app}$$

or $\qquad$ $a = F_{app}/(m_1 + m_2)$ $\qquad$ (5)

Substituting a from Eq. 5 into Eqs. 4 :

$m_1 F_{app}/(m_1 + m_2) = F_{app} - F_{2\ on\ 1}$ $\quad$ or

$F_{2\ on\ 1} = F_{app} - m_1 F_{app}/(m_1 + m_2) = F_{app}(m_1 + m_2)/ (m_1 + m_2) - F_{app}m_1/ (m_1 + m_2)$

$\qquad$ $F_{2\ on\ 1} = F_{app}\ m_2/(m_1 + m_2)$ $\qquad$ (7)

For the other Eq. 4

$\qquad$ $F_{1\ on\ 2} = m_2\ a = F_{app}\ m_2/(m_1 + m_2)$ $\qquad$ (8)

Thus, the Newton's Third Law pair ARE equal in magnitude AND they are the <u>particular value</u> above. (This is the force required to accelerate the front block, m_2.)

In this experiment, $F_{app} = T$, the tension provided by a string. Think of the tension force, T, as applied by a monkey. Monkeys are too expensive, so we use a hanging mass.

Please beware of the possible *confusion* between masses on the table (carts) and the hanging mass. That's why a tunable force from a monkey would be more clear, conceptually.

And because this is a coupled system of masses, $T \neq m_{hanging}\ g$.

➢ *Open Capstone Workbook 7C in the Capstone Labs folder on your desktop and Journal.*

Name _____

Lab 7: Impulse: How External Forces Are *Really* Exerted

<u>Post-Lab Concept Issues</u>:

Please respond to the following questions again, this time hopefully with more detail. You may also discover that you have changed your mind on an issue: Note that below:

    1. Stopping an object depends on its velocity.

    2. According to the momentum equation, it is harder to stop an object with a bigger mass.

    3. It is important that the velocity of the object should be low to stop it.

Lab 8: Conservation of Linear Momentum

Name _____

Pre-Lab Concept Issues:

Please respond to the following questions without reading ahead in the Lab book or your Text book. Like free-association, just put down on paper what your gut-level response is:

1. Does conservation of momentum only applies to collisions?

2. Is momentum conserved in collisions with immovable objects?

3. Is momentum always conserved for any system?

4. Are all collisions elastic?

Lab 8: Conservation of Linear Momentum

Equipment:

2.0 m Dynamics track
Pasco Carts, magnetic bumper on one end, Velcro bumper on the other
Two Pasco motion detectors
Cart Launcher

Introduction:

Let's explore what an external force is.

First, there are no isolated forces. The ALWAYS come in pairs that obey Newton's Third Law.

However, it is often useful to consider an isolated system of masses, and leave outside that system one (or more) of the pairs of forces.

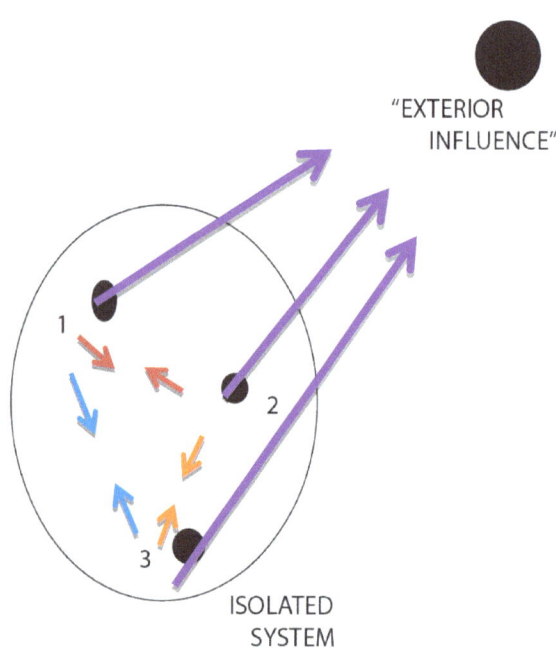

Fig 8-1 Notice the internal forces add to zero in pairs; if there is an exterior "object", then there are net forces on our Isolated System from without.

For instance, we have already been doing this with gravity. We say that if you are near the surface of the Earth, then you have weight due to the gravity of the Earth, but we left out the Earth for the sake of simplicity. Weight is the force of gravity on a mass of interest, due to another mass (often astrophysical in nature because gravity is relatively week).

Suppose, however, we have a closed system of masses with <u>NO external Forces</u>.

Thus, there are no changes in the TOTAL VECTOR MOMENTUM. Masses inside this system can exchange momentum between them, but the vector total remains the same.

Recall: "Remains the same" we call conservation.

If (and only if) there are no net external forces (you must prove this EVERY TIME before using this technique), then momentum is conserved, i.e.

$$\sum_i \vec{p}_{Ai} = \sum_i \vec{p}_{Bi} = \sum_i \vec{p}_{Ci} = \dots \quad for\ t = A, B, C \dots,\ i = 1, 2, 3 \dots \text{for the } i^{th} \text{ mass}$$

Let's restrict ourselves to two masses:

$$m_1 v_{1b} + m_2 v_{2b} = m_1 v_{1a} + m_2 v_{2a}$$

Before Collision

After Collision

As we found in Lab 8, collisions also involve energy conservation.

<u>Case 1: Completely Elastic</u>: Only kinetic energy is involved. No heat, sound, or other form of energy is involved. Actually, here on Earth, this is rarely true!

$$E_1 = K_1 = K_2 = E_2$$

Case 2: Partially Elastic: Kinetic energy and heat energy and other forms are involved which we lump into W_{other}.

$$E_1 = K_1 = K_2 + W_{other} = E_2$$

Case 3: Completely Inelastic: Kinetic energy and W_{other} are involved, with W_{other} probably the dominant effect. For this case you are looking for the phase "stick and move off together".

$$E_1 = K_1 = K_2 + W_{other} = E_2$$

$$m_1 v_{1b} + m_2 v_{2b} = (m_1 + m_2) v_{mutual}$$

Explosions: Totally Inelastic Collisions Run Backwards.

One mass fractures and moves off as an assembly of two-mass-pairs in opposite directions.

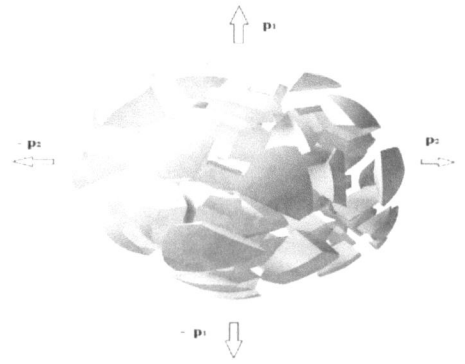

$$\Sigma\, p_i = 0 = \Sigma\, p_f$$

$$W_{other} = \Sigma\, \tfrac{1}{2}\, m_i v_i^2$$

We will be exploring these three cases, with two masses, in one-dimension.

> **Open Capstone Lab 8 and Journal.**

Name _____

Lab 8

What Happens If There Are No External forces? Momentum Conservation!

<u>Post-Lab Concept Issues</u>:

Revisit the Pre-Lab questions.

For questions where your thinking has not change, mark NC.

For questions where the experiment has changed your understanding, re-answer the question as you now understand it.

1. Does conservation of momentum only applies to collisions?

2. Is momentum conserved in collisions with immovable objects?

3. Is momentum always conserved for any system?

4. Are all collisions elastic?

Lab 9: Rotational Dynamics

Name _____

<u>Pre-Lab Concept Issues</u>:

Please respond to the following questions without reading ahead in the Lab book or your Text book. Like free-association, just put down on paper what your gut-level response is:

1. To make an extended object rotate about a given axis, only a Force is required.

2. Torque is a twisting/rotating force

3. The cause of rotational motion is force weighted by r; the inertia of a rotating system is mass weighted by r^2. Right? Isn't nature symmetric?

Lab 9: Rotational Dynamics

Equipment

Mounted rotary motion sensor with its large disk attached
Computer interface
Two objects to measure the moment of inertia, I, of an "unknown"; suggest disk, ring, or rectangular bar

Introduction

Rotating objects (at least in one dimension) satisfy Kinematics & Dynamics equations that are in exactly the *same* FORM as the translational equations, but with a different meaning of Applied Force and Mass (Inertia).

The Kinematic Definitions are similar to those for translation, except the unit **meter** becomes the unit **radian**, and we use Greek Symbols to distinguish from translation.

- Angular Position: θ [rad]
- Angular Displacement: $\Delta\theta = \theta_f - \theta_i$ [rad]
- Angular Speed: $\omega = \Delta\theta/\Delta\tau$ [rad/s]
- Angular Acceleration: $\alpha = \Delta\omega/\Delta\tau$ [rad/s^2]

The beauty of algebra is that the variables are "dummy", meaning they are place holders for whatever. Given that the Definitions are identical in form in linear motion and one dimensional rotation, we can work backwards from ⬚ to ⬚ in the same manner as for translation, and get the same results:

$$\theta = \theta_0 + \omega_0 t + 1/2\ \alpha t^2$$

$$\omega = \omega_0 + \alpha t \quad\quad\quad\quad\quad\quad\quad\quad (1)$$

$$\omega_{avg} = (\omega_i + \omega_f)/2 \quad\text{and}\quad \theta = \omega_{avg}\ t$$

$$2\ \alpha\Delta\theta = \omega_f^2 - \omega_i^2$$

Thus, the techniques for solving *these* equations are the identical to that in the first few Chapters of your text. So far, so good!

Now we need Newton's Second Law for rigid body rotation:

$$\tau = I\,\alpha \qquad\qquad\qquad \tau = r\,F\,\sin\phi$$

Notice that the FORCE has been replace by the TORQUE, and the MASS has been replaced by the ROTATIONAL INERTIA.

The source that drives the angular acceleration is an applied force, but augmented to account for the application point (relative to the rotational axis) of that force.

$$\tau = \ell\,F \qquad\qquad\qquad (3)$$

where ℓ is the lever arm, the perpendicular distance between the line of force (or action) and the axis of rotation. Computing torques is mostly geometry.

Figure 9-4: Applying a Torque.

The inertia of a rotating body depends upon the mass, but this mass is weighted by r^2, where r is the perpendicular distance between a portion of the total mass and the rotation axis. Calculating rotational inertia mostly requires calculus, or an experiment. The definition says if you divide the extended body into chunks Δm, then

$$I = \Sigma\,\Delta m_i\,r_i^2 \,/\, \Sigma\,\Delta m_i \qquad\qquad\qquad (5)$$

Fig 9-5 Same body, different axes of rotation. In the right Figure, more of the mass must rotate further from the axis so I is large. In the left Figure, more of the mass is closer to the axis and I is smaller.

Experimentally, if we wish to find the moment of Inertia, I, we can measure the torque, τ, and the angular acceleration α.

$$\tau = I\alpha \quad \text{or} \quad I = \tau/\alpha \qquad (6)$$

We have an instrument that measures rotational position, speed and acceleration. We can select to measure α and compute $\tau = F\ell$.

We apply a force of tension, F = T, through a string which is wound around a spool of radius

$R = \ell$.

➤ *Open Capstone Workbook 9 in the Capstone Labs folder on your desktop and Journal.*

Name _____

Lab 9

Rotational Motion

Post-Lab Concept Issues:

Revisit the Pre-Lab questions.

For questions where your thinking has not change, mark NC.

For questions where the experiment has changed your understanding, re-answer the question as you now understand it.

1. Stopping an object depends on its velocity. An object, which has the highest momentum, is always faster than the others.

2. According to momentum equation, it is harder to stop an object with a bigger mass.

3. It is important that the velocity of the object should be low to stop it.

4. If an unbalanced force acts on an object its momentum will change with time. Newton's Second Law states that the unbalanced force is directly proportional to the rate of change of momentum with time and is in the same direction as the force.

Lab 10: Oscillators

Name _____

Please respond to the following questions without reading ahead in the Lab book or your Text book. Like free-association, just put down on paper what your gut-level response is:

1. What does oscillation or vibration mean?

2. What condition or conditions are necessary for oscillation to occur?

3. What does resonance mean?

Lab 10: Oscillators

Equipment:

2.0 m Dynamics track
End bumpers
Screws
Motion cart
Note card or tongue depressor
Motion detector, to the side
Wave driver
String
2 springs, k = 500 N/m

Introduction:

An oscillator is a piece of matter (mass) that moves through time and space; in particular, it returns to the same spatial point every characteristic time, called the period, T.

The conditions under which a mass moves in periodic, or oscillatory motion are:

1. An equilibrium position exists. If undisturbed the mass will be in stable equilibrium (stay there forever)

2. If disturbed from equilibrium, there is a two-sided restoring force (trying to restore the object back to the equilibrium point)

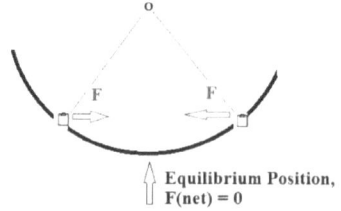

3. And our friend, **inertia**, keeps the mass in motion through the equilibrium point when the net force (≈ acceleration) is zero, and the velocity is highest.

Characteristic Measurements

- The *amplitude*, A, is the <u>maximum</u> magnitude of excursion from equilibrium. [m, degrees]

- The *frequency*, f, is the cycles (number of times the mass returns to maximum on one side) per unit time. [cycle/s = Hz]

Another way (though not independent) to express the frequency is to define the period to be

$$T = \text{time/cycle} = 1/f$$

If we graph the <u>position</u> vs. time, we find a characteristic shape: the sine curve:

These systems typically have a characteristic frequency dependent upon properties of the piece of matter: inertia and the restoring source:

$$f \sim \sqrt{\frac{Re\,storingForce}{Inertia}} \qquad (1)$$

This frequency is called the "natural" frequency.

Home Experiment 1

Find a ruler and some kind of weight, such as play-dough.

1. With the bare ruler on the table, slide increasing lengths over the edge and tweak. Listen to the frequency you hear

2. Slide the ruler over the edge of the table at a fixed length. The add mass to the end in increments.

Each time you add mass, lift the ruler, grasp at the end away from the mass, and swing in arcs in the air. How difficult is it to swing each configuration?

3. Put it back on the table with the fixed length you chose, tweak it, and listen to the frequency you hear.

Resonance

You have just discovered the principle of resonance. From Wikipedia:

"In physics, **resonance** is the tendency of a system to oscillate with greater amplitude at some frequencies than at others. Frequencies at which the response amplitude is a relative maximum are known as the system's **resonant frequencies**. At these frequencies, even small periodic driving forces can produce large amplitude oscillations, because the system continually absorbs vibrational energy." – Wikipedia

---

Home Experiment 2

Go around your domicile, and drop various (non-living, non-breakable) objects, and notice the frequency at which they "ring". The frequency you hear depends upon the factors in Eq. 1.

Some serious cases of ringing can be found in the Figures below: to the breaking point!

Figure 10-1 a) Breaking a wine glass with a true pitch voice, and b) the collapse of the Tacoma Narrows bridge.

Driven Systems

You can also try to vibrate a system externally with a wide-ranging frequencies. Again, it will absorb the most energy and vibrate with largest amplitude at its natural frequency. Example: singer breaks glass, wind breaks bridge.

Rushing air and sudden impacts produce a <u>wide</u> range of frequencies. Even these small driving forces can "pump up" a resonant system and produce large amplitude responses (as energy is stored each cycle and the total amount increases).

Example: Spring-Mass System

A spring connected to a mass satisfies all of the conditions necessary for oscillatory motion.

The formula for the force as a function of Δx (compress or stretch) is:

$$F = - k \, \Delta x$$

The minus sign means that the force OPPOSES any motion, for either sign of Δx, i.e., trying to return the system to equilibrium.

Compressed   <= F_{spring} $\Delta x < 0$

Equilibrium $F_{spring} = 0$ $\Delta x = 0$

Stretched   Fspring => $\Delta x > 0$

The **k** is called the ***spring constant***; the larger its value, the "stiffer" the spring, meaning the more difficult to compress or stretch. Units of k are N/m.

If we plot F(x), we find an ideal portion which fits the equation above, as well as the reality of ruining the spring's ability to relax to its equilibrium position by stretching it past its full restoration capability to the fracture point:

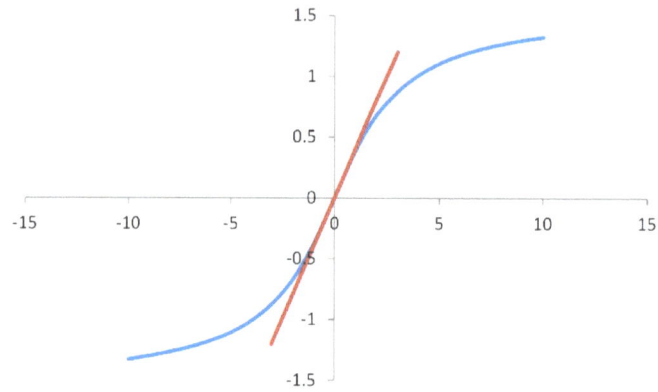

The "natural" frequency for the spring-mass oscillator is $f = \dfrac{1}{2\pi}\sqrt{m/k}$

➤ **Open Capstone Workbook 10 and follow instructions. Journal.**

Name _____

Lab 10

Oscillators

<u>Post-Lab Concept Issues</u>:

Revisit the Pre-Lab questions.

For questions where your thinking has not change, mark NC.

For questions where the experiment has changed your understanding, re-answer the question as you now understand it.

1. What does oscillation or vibration mean?

2. What condition or conditions are necessary for oscillation to occur?

3. What does resonance mean?

Lab 11: Waves

Name _____

Please respond to the following questions without reading ahead in the Lab book or your Text book. Like free-association, just put down on paper what your gut-level response is:

1. Waves transport matter.

2. Waves do not have energy.

3. Big waves travel faster than small waves in the same medium.

4. Pitch in sound is related to intensity.

Lab 11: Waves

Equipment:

Wave Driver
Dynamics track
String
Pulley
Weight hanger
200 g mass
2 paper clips
Open Sound Tube
Solid reflector for bottom end of tube (maybe a tile)
Clicker
Sound Sensor
Chladni Plates
Salt or sand

Introduction:

Recall: An OSCILLATOR is a piece of matter (mass) that moves through time and space; in particular, it returns to the same spatial point every characteristic time, called the period, T. It has an amplitude A.

A WAVE is a shape that moves through space and time. This shape moves, but it has no mass.

How can a shape, with no mass, travel? Consider a giant pair of scissors. You open the blades wide, and then begin closing them. Consider the "point" where the blades first cross in space. That "point" moves, but does not have mass. The scissors do have mass.

One view of this, is that the "point" carries information, i.e. "where the blades first cross in space and time".

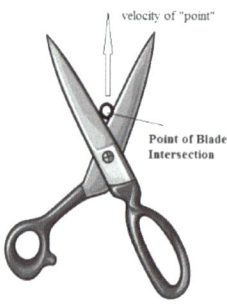

For mechanical waves (meaning, travel through matter), this shape is the highly organized, cooperative motion of a <u>large</u> number underlying oscillators.

A slinky compressional wave is shown below:

Let's look at some foam bubbles riding near the surface of a water wave.

An individual bubble rides up and down; it is an underlying oscillator.

Each bubble does this, being collectively connected by cohesion.

The "shape", however, moves horizontally. The "shape" does not contain mass, but it does carries energy. At the end where the wave exits the system, one could attach it to a piston, which would move cyclically up/down, and do work on some motor.

What can we measure to distinguish different waves?

Since we have already identified two quantities for oscillators, and since these underlying oscillators are collectively what maintains the shape, we will also associate them with the wave; in the case above, the boat is an apparent oscillator and we can measure A and f from its motion.

$$f = \text{frequency} \qquad A = \text{amplitude}$$

Notice that there are <u>two</u> spatial measurements to the waves above: the amplitude of the underlying oscillators, and the distance between crests in the "shape"; we call this the wavelength:

$$\lambda = \text{wavelength}$$

Without going through the derivation, there is also a combination of these fundamental measurements that describes an important concept: how fast does the wave move?

$$v_w = f \lambda$$

Note: this is the velocity of an insubstantial, the "shape". Think of running along a pier in pace with a wave crest. Your speed is the wave's speed.

This speed depends upon a multitude of characteristics of the matter supporting the wave, but just consider it to be a complicated constant.

Waves have a plethora of interesting behaviors (next Semester topic). We will be looking at one of these, standing waves.

Standing Waves, One Dimension

If a wave repeatedly travels and reflects from fixed end points, and if the distance between the fixed points is L, then when L is NOT an integral number of wavelengths, nothing much happens.

Below: $\lambda \neq 2L/m$ for any integer m.

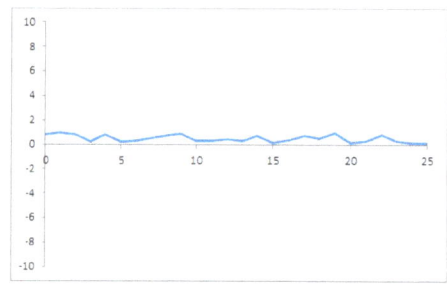

However, if there is an integer number relating the length and the wavelength

$$\mathbf{L = n\, \lambda/2} \qquad n = 1, 2, 3, 4, ...$$

(depending upon the end-point conditions) then the wave will become "resonant" and "stand" with large amplitude:

Let's look at this integer relationship between L and λ.

Below: $\lambda = 2L/n$ n = 1, 2 ,3,.....

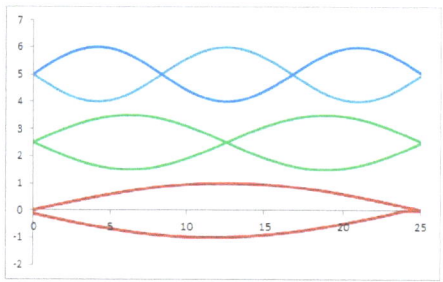

Recalling $v_w = \lambda\, f$

where v_w is a constant, then the frequencies of these configurations will be:

$f = v_w/\lambda$

and thus the longest wavelength will correspond to the lowest frequency.

We will be verifying these ideas in this lab.

> ➤ **Open Capstone Lab 11A and Journal.**

Sound Waves

Sound waves travel (in ideal air!) with a speed that only depends upon temperature:

v_{air} = [331.4 + 0.6 T(^{0} C)] m/s

We will verify this result also.

➤ **Open Capstone Lab 11B and Journal.**

Standing Waves, Two Dimensions

In one dimension, a wave travels, say to the right, encounters a boundary, and then is either reflected with the same sign (open boundary), or the opposite sign (closed boundary).

For the String waves above, both boundaries are closed.

Check your text for standing waves in pipes. The structure of the resulting standing waves changed if both end were open, both ends closed, or one end open and one end closed.

In two dimensions, the EXACT boundaries also change the standing wave structure, and change the structure in each dimension independently.

In this experiment, we will consider two different geometries in two dimensions, those on a solid flat plate. All boundaries are open, so there is an anti-node at the boundary. Actually, the resulting structure looks roughly like the tube standing waves in an tube open at both ends, and that applies in both dimensions.

1. The standing waves on the <u>rectangular plate</u> satisfy:

$L_x = n(\lambda/2)$ n = 1, 2, 3 ... $L_y = m(\lambda/2)$ m = 1, 2, 3 ...

2. The standing waves on the <u>circular plate</u> (of diameter D) have circular open edge boundaries. The calculation to produce the standing waves with these boundary conditions is a significant calculation. The "natural" coordinates for this case are polar: $\vec{r}$ = f(r,θ) and the nodes and anti-nodes lie along the radial direction, and in the azimuthal direction.

The radial nodes are determined by the zeroes of a Bessel function; the Bessel function is like a cosine function, but with decreasing amplitude at larger r. The zeroes of the Bessel function are not evenly spaced as the sine function ones are. So just observe the radial structure for conceptual enlightenment. Similarly, observe the azimuthal direction nodes and antinodes.

Procedure:

1. Lock the wave driver (see diagram). Connect a given Chladni plate to the drive shaft in each geometry. The banana plug mates directly with the hole in the drive shaft.

2. Sprinkle sand or salt on top of the plate.

3. Unlock the drive shaft of the Wave Driver.

4. Connect the Wave Driver to your function generator. Vibrate the plate over a range of frequencies from about 100 Hz up to 5 kHz. As you slowly vary the frequency of vibration, you will discover a variety of standing wave patterns. (Don't change the frequency too fast—the resonances are very sharp and you might miss some.) Adjust the amplitude and the amount of sand as necessary to get clear patterns.

5. An interesting experiment is to hold an edge of the plate to determine the effect on the resonant frequencies and patterns.

6. Repeat for the other geometry plate.

Name _____

Lab 11

Waves

Post-Lab Concept Issues:

Revisit the Pre-Lab questions.

For questions where your thinking has not change, mark NC.

For questions where the experiment has changed your understanding, re-answer the question as you now understand it.

1. Waves transport matter.

2. Waves do not have energy.

3. Big waves travel faster than small waves in the same medium.

4. Pitch in sound is related to intensity.

Lab 12: Heat Engine

Name _____

Please respond to the following questions without reading ahead in the Lab book or your Text book. Like free-association, just put down on paper what your gut-level response is:

1. Heat is a substance.

2. Heat is not energy.

3. Heat does not move.

Lab 12: Heat Engine

Equipment:

Glass syringe
Hot/cold water "baths"
Pressure Sensor
Rotary Motion Sensor
Beakers for Hot/Cold water
25-40 ml flask/tube
15 g weight

Basic Definitions – Beware! These don't match common English usage

- *HEAT = thermal energy that <u>flows</u> from a hotter substance to a cooler substance. An insubstantial. (3 mechanisms)*

- *INTERNAL ENERGY = <u>internal kinetic energy (usually random in nature)</u> that is stored in translation, vibration and possibly rotation of molecules within a bulk mass.*

- *TEMPERATURE = an <u>absolute</u> scale that measures the [kinetic energy/molecule].*

- *HEAT CAPACITY/SPECIFIC HEAT = the <u>ease</u> with which a substance gives up to/takes on internal energy from its surroundings.*

Heat & Mechanical Energy

- *Chemistry units: <u>calorie</u>, based upon 1 g water rising 1^0 C.*

- *<u>Joule</u> based on PE = mgh.*

- Joule's Experiment: how far must a mass fall (mgh) to raise 1 g water 1^0 C. See Figure.

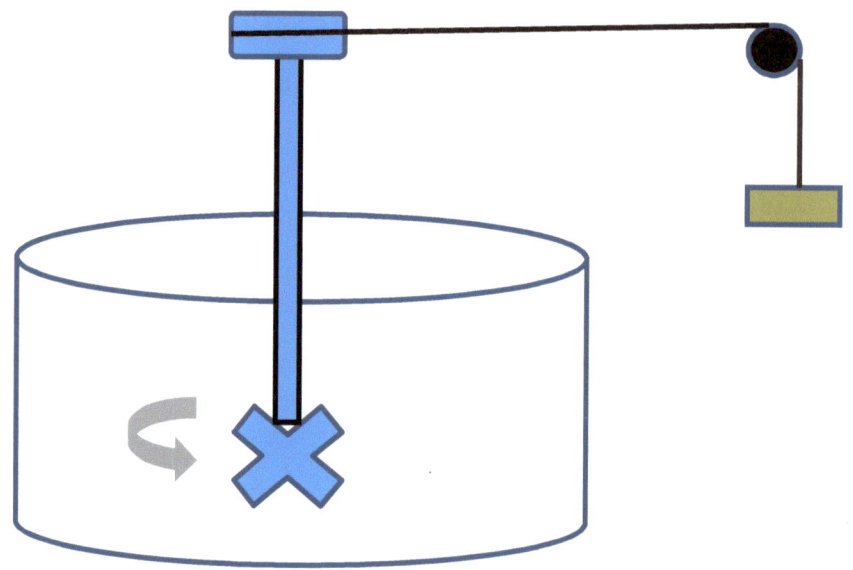

- *Result: 1 calorie = 4.186 Joule's*

- *This connects Chemistry & Physics energy units*

- *We will use the (traditional, Historical) symbol "Q" for heat energy*

Laws of Thermodynamics

- FIRST LAW: *[CONSERVATION OF ENERGY, involving U_{int}, Work, and Heat Energy Flow Q]*

- $\Delta U_{int} = Q - \Delta W$

    - *WHERE Q IS THE ENERGY <u>ADDED TO</u> THE SYSTEM*

    - *AND ΔW IS THE WORK <u>DONE BY</u> THE SYSTEM*

Heat Engines

Heat Engines operate by converting ΔU_{int} into ΔW while Q is flowing between a warm environment and a colder environment.

Ideal Model

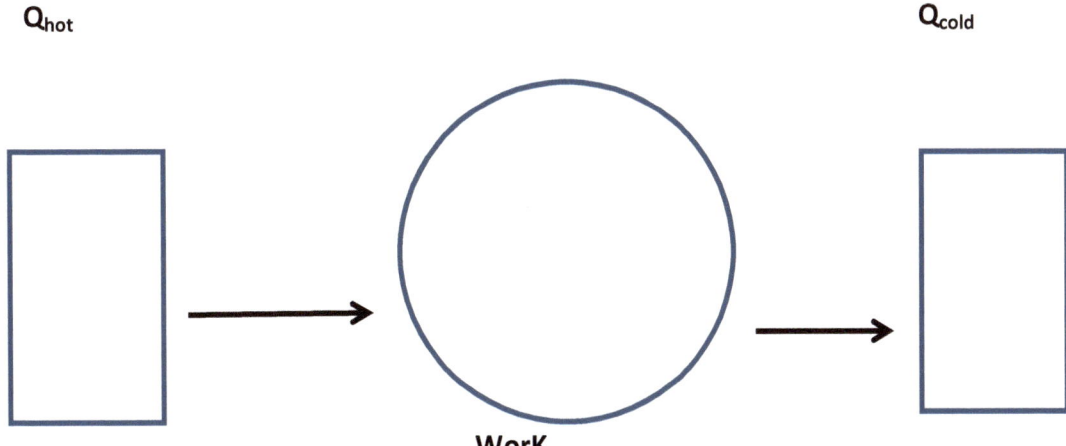

Q_{hot}   WorK   Q_{cold}

Efficiency

EFFICIENCY = work done by engine/heat maintained in its interior (which you pay for)

$$e = W/Q_{Hot} \qquad\qquad e = 1 - T_{Cold}\,(K)/T_{Hot}\,(K) \qquad (1)$$

where the latter comes from a heat capacity argument.

BEWARE, these temperatures must be expressed in Kelvin degree:

$$K = C + 273$$

Practical Device:

Fig 12-1   Water is heated, expanding steam produced, piston is moved upwards. Piston is cooled, steam contracts, piston is moved downwards.

> *Open the Link: www.youtube.com/watch?v=s3N_QJVucF8   and the figure below to see the cycle we will pursue. In the figure, adiabatic processes are possible because of thermal isolation. In our model, these portions will be done isobarically.*

This Laboratory Experiment is based upon a published paper.[19]

Aside: Practical Example or Exception?[20]

"But there is no reason why (1) should apply to engines that deliver work by a different mode of operation, according to *Janes*[21]. Indeed, the world's most universally available source of work -- the animal muscle -- presents us with a seemingly flagrant violation of that formula.

"Our muscles deliver useful work when there is no cold reservoir at hand (on a hot day the ambient temperature can be at or above body temperature) and a naive application of (1) would lead us to predict zero, or even negative efficiency. But according to *Lehninger (1965)*, under these conditions they still deliver an efficiency of about 20%.

[19] David P. Jackson, Priscilla W. Laws. Am. J. Phys. **74** (2), February 2006. Pgs. 94-101.
[20] B. Alberts *et al.,* (1983), *Molecular Biology of the Cell,* Garl a and Publishing Co., New York; pp. 550-609. E.T. Janes at http://bayes.wustl.edu/etj/articles/muscle.pdf. A. L. Lehninger (1965), *Bioenergetics,* W. A. Benjamin, N. Y. See also A. L. Lehninger (1975), *Biochemistry, The Molecular Basis of Cell Structure and Function,* Worth Publishers, Inc., 444 Park Ave. South, New York.
[21] htpp://bayes.wustl.edu/etj/articles/muscle.pdf

"According to *Alberts, et al. (1983),* under favorable conditions the efficiency of a muscle can be as high as 70%, although a Carnot engine would require an upper temperature T' of about 1000 K to achieve this.

"The answer, of course, is that a muscle is not a heat engine. It draws its energy, not from any heat reservoir, but from the activated molecules produced by a chemical reaction. That is why we should always stress the word "heat" when discussing Carnot engines.

"Only when we first allow that primary energy to degrade itself into heat at temperature T' -- and then extract only that heat for our engine -- does the Kelvin efficiency formula (1) apply. If we can learn how to capture that primary energy before it has a chance to degrade, as our muscles have already learned how to do, then we should be able to achieve higher efficiency than one would suppose from (1) in a man-made engine. Of course, this would not be a violation of the second law; rather, to achieve it will require a very clear understanding of what the second law really is."

> **Open Capstone Workbook 12 and follow directions. Journal.**

Name _____

Lab 12

Heat Engine

<u>Post-Lab Concept Issues</u>:

Revisit the Pre-Lab questions.

For questions where your thinking has not change, mark NC.

For questions where the experiment has changed your understanding, re-answer the question as you now understand it.

1. Heat is a substance.

2. Heat is not energy.

3. Heat does not move.

APPENDIX: Representative Graphs

** Note: The Capstone software allows screen-shots of all the Electronic Workbook pages, which the Student can take away in electronic form and refer to as needed.

Lab 1: Kinematics: One Dimensional Motion

Students are asked to move according to word instructions, and see what graph results; this is performed for both position and velocity. Then students are asked to view a graph (either position or velocity), match it as best as they can, and then give the word description of how they moved.

1A: Position Matching

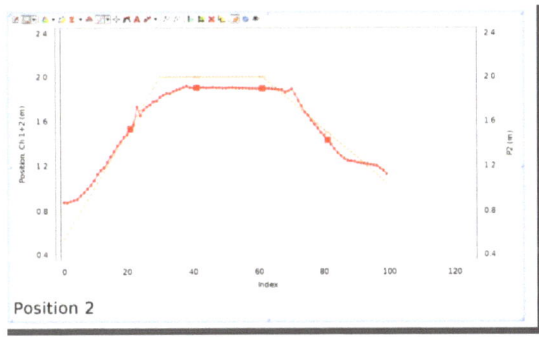

Position 2

1B: Velocity Match

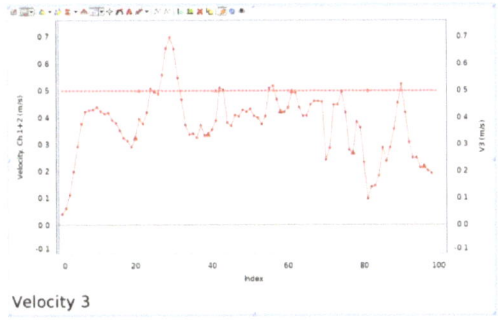

Velocity 3

Lab 2: Acceleration

Students explore both horizontal constant acceleration and vertical accelerations. Note the relationship between x, v, a in each case, and then compare across cases.

2A: Ball in free-fall (vertical)

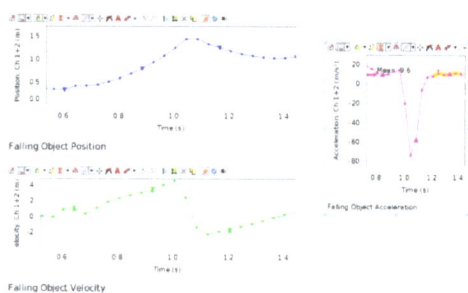

Notes: a) Look at the huge spike in acceleration when the ball touches the floor; our kinematic equations are NOT VALID at that point, b) The velocity of the ball is MAXIMUM just before it strikes the floor, c) g_{exp} = 9.6.

2B: Motion of a fan-cart (horizontal)

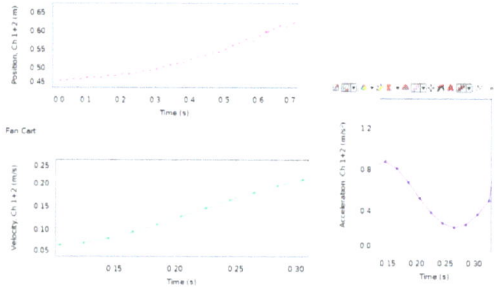

Lab 3: Projectile Motion

Students crudely measure the muzzle velocity of the projectile launcher, and then make an estimate of the range and time-of-flight. These are measured electronically for comparison. Finally, they measure the range as a function of elevation angle.

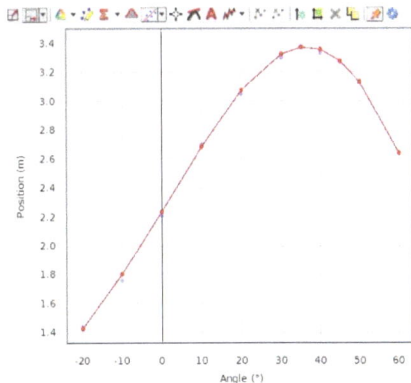

Lab 5: Adding Force Vectors & Static Equilibrium

Students use a force table. They add 1, 2, 3, and 4 forces to the central ring. They plot the graphs below. They then experimentally find the additional force that brings the system into full static equilibrium, and redraw these graphs with the additional force. Comparisons are made between graphs and experimental values.

5A: Adding Vectors tail-tail. What vector (red) will be required to bring the system of four vector forces into equilibrium?

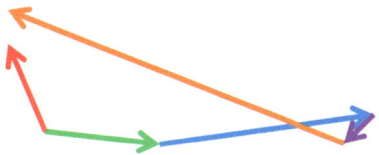

Lab 6: Conservation of Energy

6A: Hanging Weight

If there is a hanging weight supplying the tension to the dynamics cart, then the energy conversion should be

$$T \Delta x \Rightarrow \tfrac{1}{2} mv^2$$

That means the velocity (take a square root above) should be proportional to $\sqrt{\Delta x}$
The Tension should be constant
The kinetic and potential energies should be linear in Δx.

a) v(x) and T(x).

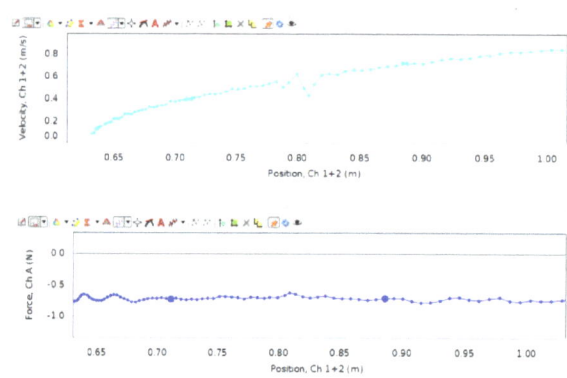

b) K, U, and E

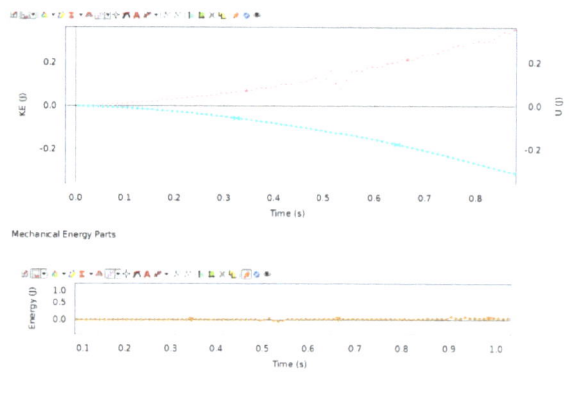

6B: Spring

If the tension is supplied by a spring, then the energy conversion should be

½ kΔx² ⇒ ½ mv²

The velocity and Tension should be directly proportional to the Δx

The kinetic and potential energies should be parabolic in Δx.

a) Spring v(x), T(x)

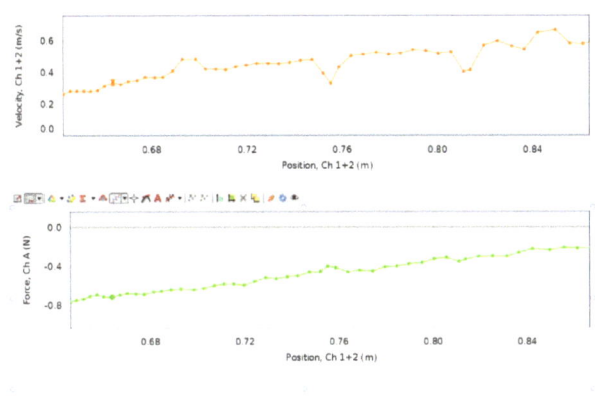

b) Spring K, U, E

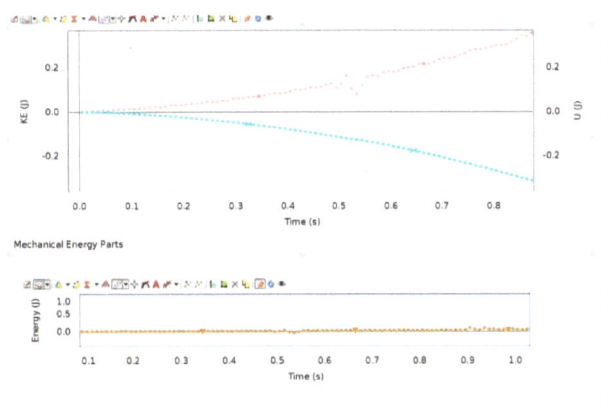

Lab 7: Impulse

A dynamics cart with a spring on the front is run into a solid barrier. The force is plotted vs. time, and the area under the curve is the impulse which should be the change in momentum. A cart launcher is used to be sure the cart always has the same initial momentum.

The same process is performed with clay on the front to make an inelastic collision. These differ by a factor of 2. Why?

Lab 8: Conservation of Linear Momentum

8A: Inelastic Collision

$m_1 = 1.0$ kg, $m_2 = 0.5$ kg

Note that at the collision point, the total kinetic energy dropped. Where did that energy go?

a) Velocities before, during, and after collision

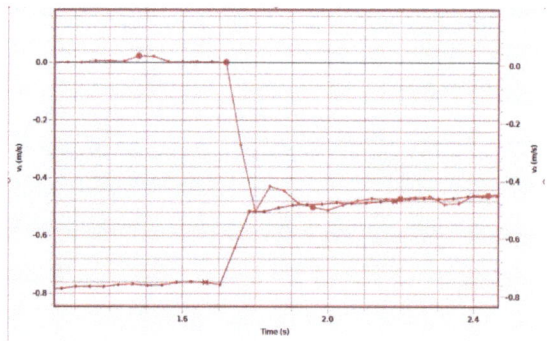

b) Momenta

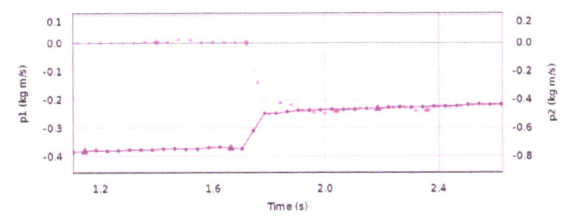

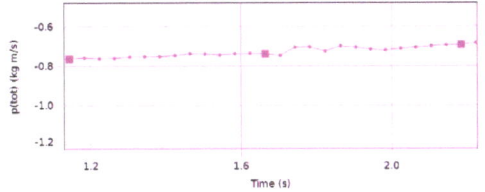

c) Kinetic and Total Energies

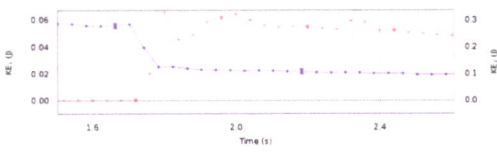

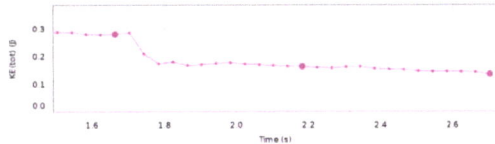

Note that the total kinetic energy dropped sharply after the collision. Where did this energy go?

8B: Elastic

$m_1 = 1.0$ kg, $m_2 = 0.5$ kg

Note that at the collision point, the total kinetic energy did not change (other than ongoing friction).

a) Velocities

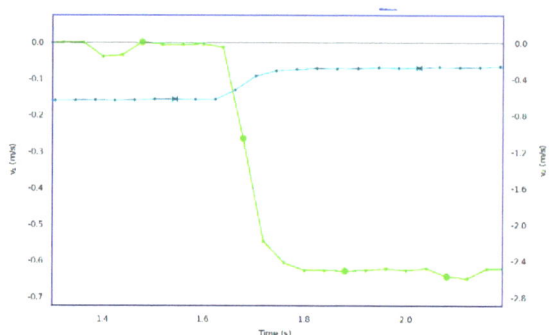

b) Momenta

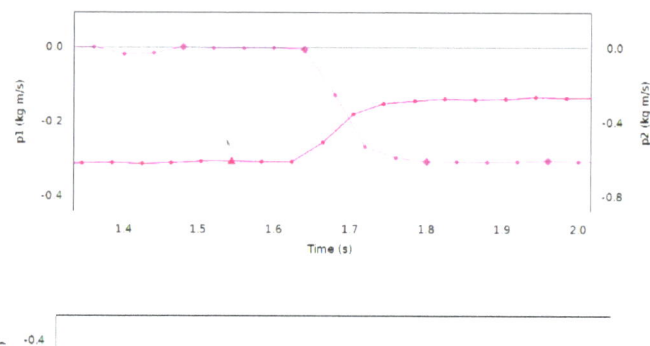

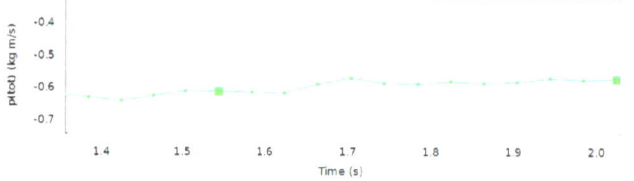

c) Kinetic and Total Energies

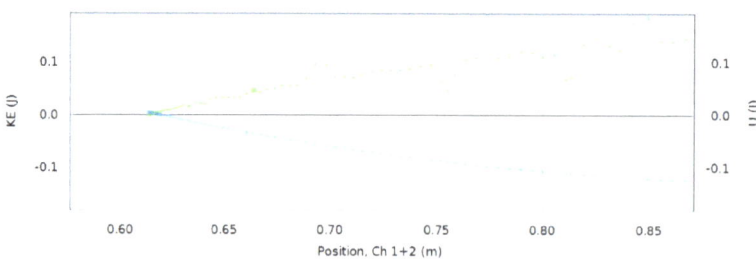

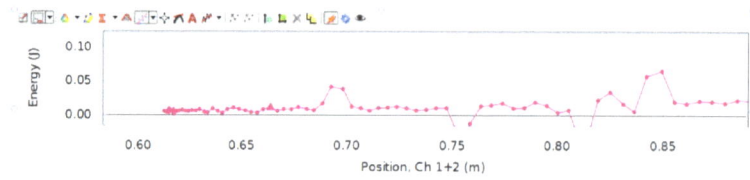

Lab 10 Oscillators

When looking at the x, v, a curves of an oscillator it is often difficult to discover the phase relationship between them. Lissajous figures are displayed for x vs. v and one finds a π/2 phase difference between them. Then an a vs. x Lissajous Figure is displayed and one finds a π phase difference between them. Students are then asked to go back to the x, v. a graphs, lay a straight edge vertically, and see if they can also detect this result.

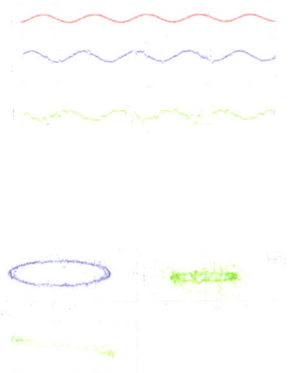

Lab 11 String Waves

Dispersion relationship and determining the string wave speed.

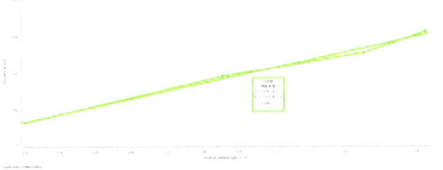

Lab 12: Heat Engine

Using a glass syringe as a piston which is connected to a small flask, the flask is placed in a hot bath, allowed to respond, and then a cold bath and allowed to respond. A test weight can be added/removed to perform work on it. The P vs. V graph for this process results.

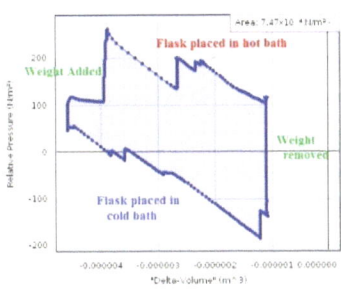

www.ingramcontent.com/pod-product-compliance
Lightning Source LLC
Chambersburg PA
CBHW050725180526
45159CB00003B/1129